SURVIVAL MATH SKILLS

Fred Pyrczak

Portland, Maine

User's Guide
to
Walch Reproducible Books

As part of our general effort to provide educational materials that are as practical and economical as possible, we have designated this publication a "reproducible book." The designation means that the purchase of the book includes purchase of the right to limited reproduction of all pages on which this symbol appears:

Here is the basic Walch policy: We grant to individual purchasers of this book the right to make sufficient copies of reproducible pages for use by all students of a single teacher. This permission is limited to a single teacher and does not apply to entire schools or school systems, so institutions purchasing the book should pass the permission on to a single teacher. Copying of the book or its parts for resale is prohibited.

Any questions regarding this policy or requests to purchase further reproduction rights should be addressed to:

Permissions Editor
J. Weston Walch, Publisher
321 Valley Street • P.O. Box 658
Portland, Maine 04104-0658

1 2 3 4 5 6 7 8 9 10

ISBN 0-8251-3819-1

J. Weston Walch, Publisher
P. O. Box 658 • Portland, Maine 04104-0658

Printed in the United States of America

Contents

*"Whole Numbers" refers to operations using whole numbers exclusively. For an exercise in which whole numbers are multiplied by proportions, for example, a dot does not appear under the "whole numbers" column.

**"Weights/Measures" refers to the *conversion* from one unit to another. The precise skills required can be determined by examining the Teacher pages for the section.

Introduction to the Third Edition

Survival math skills are those skills necessary for coping with the demands of modern society. They are used in planning adequate nutrition, arranging for transportation, maintaining savings and checking accounts, preparing household budgets, making credit purchases, understanding job benefits, paying taxes, and for a variety of other everyday purposes.

Educators have always recognized the need to teach everyday applications of math skills. Nevertheless, recent studies have indicated that many high school students and adults are seriously deficient in their ability to use math in order to cope with important everyday tasks. In response to these findings, schools have placed additional emphasis on the development of functional math skills during the past decade.

Survival Math Skills has been well received and widely used throughout the country because it provides numerous practical math problems for students to master. The purpose of this revision is to update the contents while maintaining the features that make it so popular.

The Contents makes it easy for you to integrate these exercises into your existing curriculum since the computation skills and mathematical content of each exercise are given there.

All of the eight sections begin with Teacher Pages that provide you with the computational skills, mathematical content, and procedure for each exercise, plus step-by-step solutions to all problems. With these answers, you will be able to find the exact point where a student has gone wrong in solving a problem.

For Students of Varying Abilities

In some cases, you may want to work through a problem or two with your less advanced students, and then let individuals or small groups try the remaining problems on their own.

You may also find it helpful to make transparencies to introduce exercises to students. The publisher grants to purchasers of this book the right to make transparencies of individual exercises for single classroom use.

Achievement Tests

You can monitor student progress by using the achievement tests included at the end of each section. You may want to administer the test both before and after using the section. This permits an estimate of the amount of learning that has occurred.

The tests consist of one or two questions on the material in each exercise. (Do not hold students responsible for test questions associated with exercises not assigned to them.)

Teachers who do not wish to use the tests for evaluation purposes may use them for instructional purposes. For example, solving the problems in the test may be done as an end-of-term or end-of-unit review activit{n

Section 1: Food Costs and Nutrition

TEACHER PAGES

1. Eating Out I

Computational Skills
Addition, Multiplication

Mathematical Content
Percents/Proportions/Decimals, Currency

Procedure

1. Be sure students know the meaning of the word "subtotal." In this case, it is the cost of food before tax is added.
2. Review the meanings of the words in the first three lines of the restaurant check with students.
3. All multiplication problems in this exercise yield zeros beyond the second decimal place. Point out that they should drop the last two zeros when rounding, as is done in the example.

Part I Subtotal = \$ 15.60 (\$3.90 + \$3.95 + \$5.25 + \$2.50 = \$15.60)

Tax = \$.78 (\$15.60 × .05 = \$.78)

Total = \$16.38 (\$15.60 + \$.78 = \$16.38)

Part II

1. a. \$12.60 (\$12.00 × .05 = \$.60; \$12.00 + .60 = \$12.60)
   b. \$1.80 (\$12.00 × .15 = \$1.80)
2. a. \$.59 (\$9.84 × .06 = \$.59)
   b. \$1.35 (\$9.00 × .15 = \$1.35)
3. a. \$15.60 (\$15.00 × .04 = \$.60; \$15.00 +\$.60 = \$15.60)
   b. \$3.00 (\$15.00 × .20 = \$3.00)
4. a. \$17.00 (\$14.00 + \$2.00 + \$1.00 = \$17.00)
   b. \$17.51 (\$17.00 × .03 = \$.51; \$17.00 + \$.51 = \$17.51)

2. Eating Out II

Computational Skills
Addition, Multiplication

Mathematical Content
Percents/Proportions/Decimals, Currency

Procedure

1. Discuss with the students the terms used in the menu. For example, ask them to name some different styles of eggs, ask them to describe French toast, and so forth.
2. This exercise is more advanced than the previous one since numbers other than zero appear in the third decimal place. Use Eating Out I first.
3. Name some healthy breakfast foods, and mention some benefits of having a healthy breakfast.

1. a. \$5.95 (\$4.20 + \$1.00 + \$.75 = \$5.95)
   b. \$.30 (\$5.95 × .05 = \$.2975 = \$.30)
   c. \$.89 (\$5.95 × .15 = \$.8925 = \$.89)
2. a. \$3.95 (\$2.50 + \$.75 + \$.70 = \$3.95)
   b. \$.24 (\$3.95 × .06 = \$.237 = \$.24)
   c. \$.71 (\$3.95 × .18 = \$.711 = \$.71)
3. Answers will vary.

3. Nutrition Information I

Computational Skills
Multiplication, Division

Mathematical Content
Tables/Graphs, Percents/Proportions/Decimals

Procedure

1. Discuss with students the concept of "nutritional value," that is, the worth of various food in keeping you alive and helping you grow.
2. Discuss with students the term "recommended daily allowance," that is, how much you should take in each day. Note that the labels do not tell you how much the allowance is; rather, they tell the percentage of the allowance one serving gives. Be sure that students understand that each day they should eat various foods that give them, in all, 100% of the recommended daily allowance.
3. Describe to students the following characteristics of the nutrients listed:

   protein—needed for growing muscles, bones, and other body tissues. Without this the body could not replace worn tissues.

   vitamin A—needed for growth. Lack of this causes night blindness.

   vitamin C—needed for normal metabolism, that is, the process of producing protoplasm, which is the living matter of all human cells. Lack of this causes scurvy, a disease in which gums bleed, bright spots appear on the skin, and exhaustion occurs.

   thiamine—a type of vitamin B needed by the nervous system. Lack of this causes beriberi, a disease in which arms and legs become paralyzed and the body becomes extremely thin or swollen.

   riboflavin—a type of vitamin B needed for growth. Lack of this causes stunted growth and loss of hair.

   niacin—a type of vitamin B. Lack of this causes pellagra, a disease in which the stomach and intestines become upset, the skin breaks out, and various nervous disorders occur.

   calcium—needed for healthy bones and teeth.

   iron—needed for making hemoglobin, the red part of the blood that carries oxygen.
4. Bring in a variety of foods, some healthy and some not. Have students list nutrition information for these products and compare with each other. (Use for next exercise as well.)
5. Review with less advanced students questions 1, 4, and 9 before assigning the exercise.

1. spaghetti
2. cheese
3. spaghetti
4. 60% ($4 \times 15\% = 60\%$)
5. 50% ($5 \times 10\% = 50\%$)
6. 7% ($14\% \div 2 = 7\%$)
7. 150% (5 servings; $5 \times 30\% = 150\%$)

4. Nutrition Information II

Computational Skills
Addition

Mathematical Content
Tables/Graphs, Percents/Proportions/Decimals

Procedure

1. Discuss with students the concept of a "calorie," the unit in which food energy is expressed. The body needs energy for activity and also when at rest. If you eat fewer calories than you use, you will lose weight; if you eat more than you use, you will gain weight.
2. See notes for previous exercises for descriptions of the other nutrients.
3. Review with less advanced students questions 1 and 7 before assigning the exercise.

   1. watermelon
   2. flounder
   3. peas
   4. custard
   5. a. 73% (60 + 9 + 4 = 73)
      b. 64% (4 + 60 + 0 = 64)
      c. 23% (6 + 15 + 2 = 23)

5. Sales and Coupons at the Supermarket

Computational Skills
Addition, Subtraction, Multiplication, Division

Mathematical Content
Percents/Proportions/Decimals

Procedure

1. Go over these examples with students before assigning exercises: 5¢ = $.05; 18¢ = $.18.
2. Review with less advanced students questions 1, 5, 8, and 10 before assigning the exercises. Discuss the concept of doubling coupons—the face value of coupon is doubled and then subtracted from the price of item.

   1. $.29 (2 × .89 = $1.78; $1.78 − 1.49 = .29)
   2. a. $.70 ($3.39 − 2.69 = $.70)
      b. $.45 ($2.69 ÷ 6 = $.448)
   3. $1.95 ($3.89 ÷ 2 = $1.95)
   4. $1.34 ($1.69 − .35 = $1.34)
   5. a. $7.73 ($8.48 − .75 = $7.73)
      b. Big Boy Dog Food
   6. $4.90 ($6.15 − 1.25 = $4.90)
   7. $3.19 ($2.89 + .30 = $3.19)
   8. $1.79 (2 × $.15 = $.30; $2.09 − .30 = $1.79)

Section 1 Test

1. $1.80 ($12.00 × .15 = $1.80)
2. $18.02 ($17.00 × .06 = $1.02 + $17.00 = $18.02)
3. $0.49 ($9.73 × .05 = $0.49)
4. $16.57 ($13.86 × 1.04 = $14.41 × $1.15 = $16.57)
5. 75% (25% × 3 = 75%)
6. 3 servings of spaghetti (3 × 10% = 30%)
7. 14% (10% + 4% = 14%)
8. Answers may vary but should total 50% (for example, 1 serving of watermelon, 1 serving of green peas, and 1 serving of bean soup)
9. $2.49 ($2.89 − $0.40 = $2.49)
10. Doubling coupon will make item $0.79 ($1.29 − 0.50 = $0.79)

Name ________________________________ Date ________________

1. Eating Out I

Part I

On the right is a restaurant check. Fill in the subtotal, tax, and total. The sales tax is 5% in this state.

Hint: To find 5% of a subtotal, multiply it by .05; for example, if the subtotal is $8.00, do this:

$8.00

× .05

.4000, which rounds to $.40

VALLEY GRILL

No. 1877 — Please pay cashier

Server SL — Date 9/22

No.	Item	Price	
2	Side salads	3	90
1	Pasta w/ sauce	3	95
1	Turkeyburger platter	5	25
2	Sodas	2	50
Subtotal			
Tax			
Total			

Part II

Answer the following questions.

1. a. The subtotal on Fernando's bill was $12.00. In a state with a 5% sales tax, what is Fernando's total bill? $ ____________

   b. Fernando wants to leave 15% of the subtotal as a tip. How much should he leave? $ ____________

2. a. The subtotal on Sue's bill was $9.00. In a state with a 6% sales tax, what is Sue's sales tax? $ ____________

   b. Sue wants to leave 15% of the subtotal as a tip. How much should she leave? $ ____________

3. a. The subtotal on Elise's bill was $15.00. In a state with a 4% sales tax, what is Elise's total bill? $ ____________

   b. Elise wants to leave 20% of the subtotal as a tip. How much should she leave? $ ____________

4. a. Frank ordered items costing $14.00, $2.00, and $1.00. How much is his subtotal? $ ____________

   b. In a state with a 3% sales tax, what is Frank's total bill? $ ____________

Name ______________________ Date ______________

2. Eating Out II

In the box is part of a restaurant menu. Read it and answer the questions. Round off your answer to two decimal places.

Hint: To round off, look at the third number to the right of the decimal place. If the number is 5 or larger, round up: for example, $1.875 becomes $1.88. If the number is less than 5, round down: for example, $.524 becomes $.52.

Breakfasts

Two Eggs, Any Style	2.50
Served with Toast, Butter, and Jelly	
With Grilled Ham, Crisp Bacon, or Link Sausage	3.75
Golden Wheat Cakes, Waffles, or French Toast	2.75
With Crisp Bacon, Link Sausage, or Grilled Ham	4.00
Three Egg Omelette	2.95
Served with Toast, Butter, and Jelly	
With Grilled Ham, Bacon, or Sausage	4.20
Cholesterol-Free Egg Substitute	add'l .75

Beverages

Coffee	.75
Tea	.70
Milk	.70

Chilled Juices

All Your Favorites

Orange, Tomato, Grapefruit	1.00
Large Glass	1.50

1. a. Elizabeth ordered an omelette with bacon, a small glass of orange juice, and a cup of coffee. What is the total cost of her food? $ ____________

   b. There is a 5% sales tax on restaurant food in the state. How much will the sales tax be? $ ____________

   c. Elizabeth wants to leave 15% of the subtotal as a tip. How much should she leave? $ ____________

2. a. Joe ordered scrambled eggs with egg substitute and a cup of tea. What is the total cost of his food? $ ____________

   b. There is a 6% sales tax on restaurant food in the state. How much will the sales tax be? $ ____________

   c. Joe wants to leave 18% of the subtotal as a tip. How much should he leave? $ ____________

3. a. Suppose that you could spend up to $6.00 on breakfast, including a 15% tip based on your bill's subtotal and 5% tax. Write what you would order here. ____________

   ____

   b. What is the total cost of what you ordered including tax and tip? $ ____________

Name ______________________________ Date ______________

3. Nutrition Information I

Many foods come with a label that tells about their nutritional value. Read the labels in the boxes below and answer the questions.

A package of cheese with five slices in it has this information on the label:

Serving size: one slice
Percentage of U.S.
Recommended Daily Allowance

Nutrient	%
Protein	10
Riboflavin	4
Vitamin A	4
Niacin	*
Vitamin C	*
Calcium	15
Thiamine	*
Iron	*

A 280-gram package of dry spaghetti with no sauce has this information on the label:

Serving size: 56 grams
(2ounces dry)
Percentage of U.S.
Recommended Daily Allowance

Nutrient	%
Protein	14
Riboflavin	10
Vitamin A	*
Niacin	15
Vitamin C	*
Calcium	*
Thiamine	30
Iron	10

*Contains less than 2% U.S. recommended daily allowance of these nutrients.

1. Will one serving of spaghetti or one slice of cheese give you more protein? ____________
2. Is spaghetti or cheese a better source of calcium? ____________
3. Is the spaghetti or cheese a better source of iron? ____________
4. If you ate four slices of cheese, what percentage of your daily need for calcium would you get?

   ____

5. If you ate the whole package of cheese, what percentage of your daily need for protein would you get?

   ____

6. If you ate one half serving of spaghetti, what percentage of your daily need for protein would you get?

   ____

7. If you ate the whole package of spaghetti, what percentage of your daily need for thiamine would you get? ____________

Name ______________________ Date ______________

4. Nutrition Information II

Below is nutrition information on some foods. In the first column are the names of foods. In the second column is an amount of each food that you might eat in a meal. In the third column is the number of calories you would get by eating that amount of food. In the last eight columns are the percentages of various nutrients you would receive by eating each food. For example, if you ate one slice of rye bread, you would receive 4% of your need for protein for that day. Where you see an asterisk (*), count it as 0% in working out your answers.

Food energy and percentage of U.S. Recommended Daily Allowance for eight nutrients provided by a specified amount of various foods

Food	Amount	Food energy	Protein	Vitamin A	Vitamin C	Thiamine	Riboflavin	Niacin	Calcium	Iron
		Calories	*Percentage of U.S. RDA*							
Rye bread: light, 18 slices per 1-lb loaf	1 slice	60	4	*	*	4	2	2	2	2
Flounder: baked or broiled	112 grams (4 ounces)	133	60	*	4	4	4	10	2	6
Watermelon: raw diced pieces	1 cup	50	2	20	20	4	2	2	2	4
Peas, green: cooked, drained	1 cup	134	9	15	60	30	10	20	4	15
Soups: canned, condensed										
Bean with pork	1 cup	240	10	15	4	8	4	4	6	10
Beef broth	1 cup	30	8	*	*	*	2	6	*	2
Desserts: custard, baked	1 cup	310	30	20	-	8	30	2	30	6
*None or less than 1 percent										

1. Which sweet food has the least number of calories? ______________

2. Which food is the best source of protein? ______________

3. Which food is the best source of iron? ______________

4. Which food is the best source of riboflavin? ______________

5. a. If you ate flounder, peas, and rye bread in the amounts given in the box, what percent of your daily need for protein would you receive? ______________

   b. What percent of your daily need for vitamin C would you receive? ______________

   c. What percent of your daily need for iron would you receive? ______________

Name ______________________________ Date ______________

5. Sales and Coupons at the Supermarket

Below are some questions on prices you might pay at a supermarket. Answer the questions in the spaces. For problems requiring division, complete your answers to three decimal places and round off to two.

1. The sale price for lettuce is two heads for $1.49. The regular price is 89 cents per head. How much do you save by buying two heads at the sale price instead of two heads at the regular price? $ __________

2. a. The sale price for a six-pack of soda is $2.69. The regular price is $3.39. How much do you save by buying a six-pack at the sale price instead of the regular price? $ __________

   b. How much does each can of soda cost at the sale price? $ __________

3. The sale price for a package of two lightbulbs is $3.89. How much does each lightbulb cost you at the sale price? $ __________

4. You have a coupon that says "35 cents off on any bottle of Good Home Salad Dressing." A bottle regularly costs $1.69. How much will it cost you with the coupon? $ __________

5. a. You have a coupon that says "75 cents off on a large-size bag of Big Boy Dog Food." A large bag usually costs $8.48. How much will it cost you with the coupon? $ __________

   b. You usually buy XYZ Dog Food. A large-size bag regularly sells for $7.95. It is *not* on sale. If both brands are equally good, should you buy Big Boy Dog Food on sale or XYZ at the regular price?

   ____

6. The label on a bottle of car wash says "Mail in attached coupon with your cash register receipt and receive a $1.25 refund." The bottle sells for $6.15. Considering the refund, how much is your final cost? $ __________

7. A tube of toothpaste has these words on it: "Price marked is 30 cents off the regular price." The price on the box is $2.89. What is the regular price? $ __________

8. A supermarket is advertising that it will give you *double* the amount of your coupons off the selling price. You have a coupon for 15 cents off a Tasty Toothbrush. The selling price is $2.09. How much will the toothbrush cost you if you use your coupon at the supermarket giving double savings?

   $ __________

Name ______________________________ Date ______________

Section 1: Food Costs and Nutrition Test

1. Maria ordered $12.00 worth of food in a restaurant. She wants to leave a 15% tip. How much should she leave as a tip? $ __________

2. Charlene ordered $17.00 worth of take-out food at her local Chinese restaurant. If she paid 6% sales tax on her order, what was the amount of her total bill? $ __________

3. Miguel ordered a $9.73 dinner in a restaurant. In his state there is a 5% sales tax on restaurant food. How much is the tax? $ __________

4. If you were to leave a 15% tip on the total cost of a meal subtotaling $13.86 and sales taxed at 4%, what is the total amount you would spend on this order? $ __________

5. If one slice of cheese gives you 25% of your daily need for calcium, what percent of your daily need would three slices give you? __________

6. A serving of spaghetti provides 10% of your daily need for iron, and a serving of flounder provides you with 6%. Which would give you a greater daily intake of iron: three servings of spaghetti or four servings of flounder? __________

7. If a serving of bean with pork soup gives you 10% of your daily need for protein and a slice of rye bread gives you 4%, what percent will you have taken care of if you eat both?

   ____

8. A serving of green peas and a serving of bean soup each provides 15% of your daily requirement of vitamin A. A serving of baked custard and a cup of diced watermelon each provide 20%. What combination of these foods can you eat to obtain 50% of your daily requirement of vitamin A? ______________________________

   ____

9. You have a coupon that gives you 40 cents off on a purchase of window cleaner. The window cleaner regularly costs $2.89. How much will it cost you after you use the coupon? $ __________

10. Which of the following is a better deal: doubling a 25-cent coupon and using it toward a name-brand item selling for $1.29 or buying a store brand of the same item that sells for 89 cents? ______________________________

Section 2: Transportation and Vacation Costs

TEACHER PAGES

6. Costs of Travel by Auto

Computational Skills
Subtraction

Mathematical Content
Tables/Graphs, Whole Numbers, Currency

Procedure

1. Review with less advanced students questions 1, 5, and 7 before assigning the exercise.
2. Talk about where the students can find a gas mileage guide and how engine size, city and highway mileage, and cost per gallon all affect estimated yearly cost for gasoline.

1. 22
2. 20
3. 3 (26 – 23 = 3)
4. \$758
5. \$314 (\$847 – \$533 = \$314)
6. \$42 (\$655 – \$613 = \$42)

7. Costs of Buying Used Autos

Computational Skills
Subtraction, Division

Mathematical Content
Whole Numbers, Currency

Procedure

1. Review with less advanced students questions 1a through 1e before assigning the exercise. Make sure students understand all the new terms before proceeding with the problems. Can students list other features cars and trucks have that make them more valuable to the consumer?

1. a \$499.00
   b. \$155.00
   c. 36 months
   d. 3 years (36 ÷ 12 = 3)
   e. \$885 (\$6,079 – \$5,194 = \$885)
2. \$899 (\$4,715 – \$3,816 = \$899)
3. \$70.00 (\$162.00 – \$92.00 = \$70.00)

8. Costs of Buying Vans, Motorcycles, and Cars

Computational Skills
Addition, Subtraction, Multiplication

Mathematical Content
Percents/Proportions/Decimals, Currency

Procedure

1. Discuss factors that may take away from the value of a vehicle—condition, number of miles, extra features, year and model of car, etc.
2. Go over the sample computation given in the Hint.

1. \$2,007 (\$8,995 – \$6,988 = \$2,007)
2. \$7,950 (\$10,900 – \$2,950 = \$7,950)
3. \$1,595 (\$10,995 – \$9,400 = \$1,595)
4. \$95 (\$10,995 – \$10,900 = \$95)
5. a. \$307.52 (\$7,688 × .04 = \$307.52)
   b. \$7,995.52 (\$7,688 + \$307.52 = \$7,995.52)
   c. \$76.88 (\$7,688 × .01 = \$76.88)
   d. \$8,072.40 (\$7,688 + \$307.52 + \$76.88 = \$8,072.40)
6. \$4,297.50 (\$4,775 × .10 = \$477.50; \$4,775 – \$477.50 = \$4,297.50)

9. Car Leasing

Computational Skills

Addition, Multiplication

Mathematical Content

Tables/Graphs, Percents/Proportions/Decimals, Currency

Procedure

1. Make sure students understand what the following terms refer to:

   MSRP—Manufacturers Suggested Retail Price, the cost the maker suggests the dealer sell the car for

   HP engine—amount of horsepower the engine has

   dual air bags—air bags for driver and passenger, a safety feature

   ABS brakes—antilock brakes, a safety feature

   MVD—motor vehicle department, which charges fees for licensing, registration, plates

   security deposit—paid up front when leasing, refundable when vehicle returned at end of lease term, provided that the vehicle is in good condition

   Down Payment—up-front payment in addition to the monthly charge
2. Discuss that, when leasing, it is the owner's responsibility to take proper care of the leased vehicle or additional fees may be charged when the vehicle is returned. Maintenance fees are paid by the person leasing the vehicle.
3. Discuss the benefits of leasing a vehicle.
4. Discuss the benefits of owning a vehicle.
5. Remind students that any down payment should be added to the monthly cost to determine the full amount being charged; it is an additional cost.
6. Point out the extra fee associated with driving a leased vehicle more miles than agreed upon in a lease contract. A review of multiplication by decimals may be helpful.

1. a. \$259
   b. \$217
2. a. \$3,108 (\$259 × 12 = \$3,108)
   b. \$2,604 (\$217 × 12 = \$2,604)
3. \$9,324 (\$259 × 36=\$9,324)
4. \$9,012 (\$217 × 36 = \$7,812 + \$1,200 = \$9,012)
5. \$200 (\$0.10 × 2,000 = \$200)

10. Travel by Bus

Computational Skills
Addition, Subtraction

Mathematical Content
Tables/Graphs, Weights/Measures

Procedure

Before assigning this exercise to less advanced students, discuss schedules 3501 and 3502, neither of which provides answers to the questions in this exercise. For schedule 3501, for example, you might ask:

1. Is this bus going to San Francisco or away from San Francisco? (away)
2. Does it leave San Francisco in the morning or afternoon? (morning)
3. What time does it leave San Francisco? (10:30 A.M.)
4. Does it arrive in Saratoga in the morning or afternoon? (afternoon)
5. How long is the trip from San Francisco to Saratoga? (1 hour 48 minutes)

1. 3516
2. 1 hour 5 minutes
3. 3517
4. 1 hour
5. 7:20 P.M. (Note: He had to take 3495, a later bus than usual.)
6. 2 hours 5 minutes
7. 7:35
8. P.M.
9. 9 hours 40 minutes (2 hours 5 minutes until noon plus 7 hours 35 minutes after noon)
10. 3494
11. 11 hours 46 minutes (8 hours 15 minutes until noon plus 3 hours 31 minutes after noon)
12. 8 hours 10 minutes (Note: From 7:35 P.M. on Saturday until midnight = 4 hours 25 minutes and from midnight until 3:45 A.M. = 3 hours 45 minutes.)

11. Travel by Air

Computational Skills
Addition, Subtraction

Mathematical Content
Tables/Graphs, Currency, Weights/Measures

Procedure

1. Discuss with students the difference between daylight time and standard time. In the spring, we turn our clocks forward one hour to go from standard to daylight time. In the fall, we turn our clocks back one hour to go from daylight to standard time.

2. Have more advanced students compute the flight time for flight 22/60 from Austin to Denver (3 hours and 5 minutes) and for flight 228/28 from Austin to New York (5 hours and 10 minutes). This may be computed by first converting the arrival time to the equivalent in Central Daylight Time. The difference between the departure and arrival times, both expressed as CDT, is the answer to each question. Note that flight time in these cases may include stops at one or more cities before arriving at the destinations.
3. Review with less advanced students the second row in the box before assigning Part I.
4. Review with less advanced students questions 1, 3, and 5 before assigning Part II.

Part I

Pacific Daylight Time (PDT) (Los Angeles)	Mountain Daylight Time (MDT) (Denver)	Central Daylight Time (CDT) (Chicago)	Eastern Daylight Time (EDT) (Philadelphia)
6 A.M.	7 A.M.	8 A.M.	9 A.M.
8 A.M	9 A.M.	10 A.M.	11 A.M.
11 A.M.	12 noon	1 P.M.	2 P.M.

Part II

1. Flight 228/204
2. Flight 18
3. 4:05 P.M. (Note: you would arrive at 3:05 Denver time, which is Mountain Daylight Time. It is one hour later in Austin, which is on Central Daylight Time.)
4. 2:30 P.M. (Note: You would arrive at 3:30 New York time, which is Eastern Daylight Time. It is one hour earlier in Austin, which is on Central Daylight Time.)
5. $200 ($390 – $190 = $200)
6. $85 ($175 – $90 = $85)
7. $84 ($169 – $85 = $84)

12. Motel Costs

Computational Skills

Addition, Subtraction, Multiplication

Mathematical Content

Tables/Graphs, Currency

Procedure

Discuss each of these terms with students: in-season (time of year when many people want to stay in the area); 34 units (number of separate rooms or suites of rooms); nicely appointed; A/C (air-conditioned); direct-dial phones (can dial an outside number without going through the motel operator); movie rentals (can rent movies from motel rooms equipped with a video machine); all major cc (can pay with a credit card); pets/no pets (can or cannot bring pets to motel).

1. a. \$45 to \$50
   b. \$135 to \$150
2. a. \$99 to \$130
   b. \$198 to \$260
3. a. \$65 to \$70
   b. \$260 to \$280
4. \$8 to \$14 (\$28 – \$20 = \$8; \$36 – \$22 = \$14) \$34 to \$60 (\$79 – \$45 = \$34; \$110 – \$50 = \$60)
5. \$100 to \$110 (\$45 + \$55 = \$100; \$50 + \$60 = \$110)
6. a. \$24.00
   b. \$8.95
7. \$63.80 (\$15.95 × 4 = \$63.80)
8. \$36 (\$12 × 3 = \$36)
9. Royal Motel (120 units)
10. Royal Motel (24-hour restaurant)

13. Camping Costs

Computational Skills

Addition, Subtraction, Multiplication, Division

Mathematical Content

Tables/Graphs, Percents/Proportions/Decimals, Currency

Procedure

1. Be sure students know the meanings of these terms and abbreviations: per night fee (cost for each night); hookup vs. tent (differences in services); coin-op services (uses coins to operate those facilities).
2. Inform students that the cost for two people is the cost for either one or two campers (some like to camp alone and the rate for two would still apply to them).
   1. a. \$54.00 (\$18 × 3 = \$54)
      b. \$36.00 (\$12 × 3 = \$36)
   2. a. \$27.00 (\$18 + \$3 × 3 = \$27)
      b. \$16.50 (\$1.50 × 3 = \$4.50 + \$12.00 = \$16.50)
   3. Answers may vary, e.g.;
      a. services included, more luxurious, water available
      b. cheaper, more "camplike," extra services available if needed.

4. \$124.25 (\$16.50 × 7 = \$115.50; \$1.25 × 7 = \$8.75; \$115.50 + \$8.75 = \$124.25)
5. \$34.00 (\$1.25 × 4 = \$5.00 + \$1.50 × 2 = \$5.00 + \$3.00 = \$8.00; \$16.50 × 4 = \$66.00 + \$8.00 = \$74.00; \$27.00 × 4 = \$108.00 – \$74.00 = \$34.00)

Section 2 Test

1. \$351 (\$819 – \$468 = \$351)
2. \$657.66 (\$5,445.66 – \$4,788.00 = \$657.66)
3. 42 months (\$12,122 – \$2,000 = \$10,122 ÷ \$241 = 42)
4. \$213.48 (\$3,558 × .06 = \$213.48)
5. \$7,495 (\$186 × 36 = \$6,696 + \$799 = \$7,495)
6. 8 hours 20 minutes
7. 9 A.M.
8. \$166 (\$326 – \$160 = \$166)
9. \$222 (\$74 × 3 = \$222)
10. \$57 (\$1.75 × 3 = \$5.25 + \$9 = \$14.25 × 4 = \$57)

Name ______________________________ Date ____________________

6. Costs of Travel by Auto

In the box is part of a table showing the fuel costs for various makes of cars. You can find complete data in the *Gas Mileage Guide*, a reference published by the U.S. government and obtainable from any new car dealer. Read the table and answer the questions.

Hint: Numbers separated by a slash (/) give two different pieces of information. For example, the first numbers for a Cherokee are 5.2/8. The 5.2 is the engine size in liters; the 8 is the number of cylinders.

Manufacturer	Vehicle Description		Fuel Economy		
Model	Engine/size liters/cylinders	Transmission A= automatic M=manual	City miles per gallon	Highway miles per gallon	Average annual fuel costs
Jeep Grand Cherokee	5.2/8 4.0/6	A A	15 17	20 21	$823 $758
Eagle Vision	3.3/6 3.5/6	A A	21 19	26 25	$613 $655
Honda Accord	2.4/4 3.0/6	M A	22 18	26 23	$720 $702
Acura NSX	3.2/8	A/M	14	20	$847
Saturn	2.0/4	M	25	29	$533

1. How many miles can you travel on a gallon of gasoline in the city with a manual transmission Accord? ____________
2. How many miles can you travel on a gallon of gasoline on the highway with an 8-cylinder Grand Cherokee? ____________
3. How many more miles can you travel on a gallon of gasoline on the highway if you buy a manual transmission Accord instead of an automatic? ____________
4. How much will gasoline probably cost you per year if you buy the 6-cylinder Grand Cherokee? $ ____________
5. How much more will gasoline probably cost you per year if you buy the Acura instead of the Saturn? $ ____________
6. How much less will gasoline probably cost you per year if you buy the 3.3-liter Vision instead of the 3.5-liter Vision? $ ____________

Name ________________________________ Date ______________

7. Costs of Buying Used Autos

In the box are the advertisements for three used cars. Read them and answer the questions.

Terms to Understand: total down (total amount of cash you must pay at the time you buy the car if you are going to take out a loan for the rest of the cost); mos. (months); cash price (total price if you are going to pay cash and not take out a loan); deferred price (price if you make a down payment and take out a loan for the rest of the cost); finance charges (charges for borrowing money).

Total down	Car	Per month
$499 TOTAL DOWN	Chrysler LeBaron . . . .$4900 Factory A/C. Economy 6 cyl., 4 dr automatic, AM/FM cassette, power steering, heater, vinyl top, body side moldings, w/wall tires, wheels covers, tinted glass, accent stripes, 40K miles For 36 mos. Cash price $5194.00 includes sales tax Deferred price $6079 includes finance charges	**$155** PER MO.
$249 TOTAL DOWN	Hyundai Sonata . . . . $6995 4 dr, 6 cyl, A/C, auto, drivers side airbag, power steering, power locks, power windows, AM/FM cassette w/ CD hookup, heater, body side molding, w/wall tires, wheel covers, 34K miles For 48 mos. Cash price $7414.70 includes sales tax Deferred price $8025 includes finance charges	**$162** PER MO.
$299 TOTAL DOWN	Toyota Corolla . . . . $3600 2 dr, factory air, power steering, power brakes, automatic trans, driver-side air bag, AM/FM radio, heater, power locks, tinted glass, w/wall tires, wheel covers, body side molding, 60K miles For 48 mos. Cash price $3816.00 includes sales tax Deferred price $4715 includes finance charges	**$92** PER MO.

1. a. If you wanted to take out a loan to buy the Chrysler, how much cash would you need to have as a down payment? $ ______________

   b. How much would the payments be each month? $ ______________

   c. For how many months would you make the payments? ______________

   d. For how many years would you make the payments? ______________

   e. How much would you save if you paid the full price in cash instead of taking out a loan and paying the deferred price? $ ______________

2. How much is the difference between the cash price and the deferred price for the Toyota? $ __________

3. How much more per month would the Hyundai cost than the Toyota? $ ______________

Name ______________________________ Date ______________

8. Costs of Buying Vans, Motorcycles, and Cars

The boxes show advertisements for used vehicles. Read them and answer the questions.

Hint: To take 6% of a number, multiply it by .06.

VANS

Ford Aerostar	$10,995
7 passenger, 6 cyl, air 48K	
Dodge Caravan	$15,495
auto, air, 6 cyl, tilt, PDL, cruise, cassette, 50K	
Plymouth Grand Voyager	$9,400
auto, AC, V6, like new 56K	
VAN CONVERSION SALES	

MOTORCYCLES

Yamaha VMAX	$10,900
900 miles	
Suzuki	$2,950
extras, nice	
Honda CBR 600 F2	$3,800
dependable	
Harley Sportster	$4,775
like new, xtra chrome, 42K	
PASADENA MOTORCYCLES	

CARS

Dodge NEON (1JOE929)	**$6,988**
Chevy S-1 (2F1710 SOLD	**$5,488**
Ford TAURUS (2AWY042)	**$8,888**
Chev. CAVALIER (1LWM597)	**$SAVE$**
Toyota PIC (2R45749) SOLD	**$SAVE$**
Nissan ALTIMA (1JDB238)	**$8,995**
Pontiac SUNBIRD (1KHY968)	**$7,688**
RELIABLE MOTORS	

1. The Nissan Altima costs how much more than the Dodge Neon? $ ____________

2. The Yamaha costs how much more than the Suzuki? ____________

3. The Plymouth Grand Voyager costs how much less than the Ford Aerostar? $ ____________

4. The most expensive van costs how much more than the most expensive motorcycle? $ ____________

5. a. In a state with a 4% sales tax, what is the amount of the sales tax on the Pontiac? $ ____________

   b. What would be the cost of the Pontiac including sales tax? $ ____________

   c. In a state with a registration fee of 1% of the cost before tax, what is the amount of the registration fee on the Pontiac? $ ____________

   d. What is the cost of the Pontiac including sales tax and registration fee? $ ____________

6. The dealer at Pasadena Motorcycles is willing to give a 10% discount to anyone who is ready to pay cash. What will be the cost for the Harley Sportster if cash is paid? $ ____________

Name ______________________________ Date ______________

9. Car Leasing

Donnie plans to lease a car. He has narrowed down his choices to the two vehicles presented in the table below. Refer to the table to help answer the following questions.

Subaru Outback Wagon All-Wheel Drive, 165 HP Engine, A/C, Dual Air Bags, 80 Watt Sound System, Power Windows & Door Locks, ABS Brakes, + Much More MSRP $22,995 Lease for only: **$259 / mo.*** 36 months No $ Down
Subaru Impreza Outback All-Wheel Drive, A/C, Dual Air Bags, Cassette/CD, Power Moon Roof, Power Windows, Roof Rack, Tilt, Rear Window Defroster, + More MSRP $20,550 Lease for only: **$217 / mo.*** 36 months $1,200 down
* Prices include all rebates and discounts. Taxes and MVD fees extra. Lease for 12,000 miles/year. $0.10 charge each mile over. Total due at signing – down pmt (if any), bank fee, 1st month pmt, security deposit, and conveyance fee. Special finance available. See deal for details.

1. a. What is the monthly payment when leasing the Outback Wagon? $ ____________

   b. What is the monthly payment when leasing the Impreza? $ ____________

2. a. If Donnie leases the Outback Wagon, what would it cost him per year in monthly finance charges?

   $ ____________

   b. If Donnie leases the Impreza, what would it cost him per year in monthly finance charges?

   $ ____________

3. What is the total cost to lease the Outback Wagon for 36 months (not including taxes and other fees)?

   $ ____________

4. What is the total cost, minus taxes and other fees, to lease the Impreza for three years?

   $ ____________

5. If Donnie typically drives 14,000 miles each year, what additional amount would he owe per year if he leased either vehicle? $ ____________

Name ______________________________ Date ______________

10. Travel by Bus

In the box is a bus schedule. Read it and answer the questions.

Hint: In the columns on the left, the arrows point down, showing that these buses leave San Francisco for other cities. In the columns on the right, the arrows point up, showing that the buses are going to San Francisco.

SAN FRANCISCO—SANTA CRUZ—SALINAS

READ DOWN						READ UP					
3495	3517	3493	3501	3491	←SCHEDULE Nos.→	3516	3502	3492	3494	3496	3504
5 15	**4 45**	**1 45**	10 30	7 15	Lv SAN FRANCISCO, CAL Ar ... Ar	7 35	9 10	11 30	**4 40**	8 15	**10 25**
↓	↓	↓	10 50	↓	San Fran, Int'l Airport	↑		↑		↑	↑
		2 30	↓		Moffett Field ...		8 15	10 43	**3 37**		**9 38**
		↓	11 10	7 55	Redwood City ...		↑	↑	↑	7 35	↑
			11 30	8 10	Palo Alto ...					7 20	
			11 45	8 25	Mountain View ...					7 05	
6 10	**5 45**	**2 36**	11 55	8 35	Sunnyvale ...	6 30	8 09	10 34	**3 31**	6 55	**9 32**
6 25	**6 00**	**2 48**	**12 07**	8 48	Cupertino ...	6 20	7 58	10 24	**3 17**	6 45	**9 23**
6 35	**6 10**	**3 00**	**12 18**	8 59	Saratoga ...	6 12	7 50	10 15	**3 08**	6 35	**9 17**
6 45	**6 20**	**3 12**	**12 30**	9 10	**Los Gatos** ...	6 05	7 41	10 08	**3 00**	6 25	**9 10**
7 10		**3 35**	**12 55**	9 35	Scotts Valley ...		7 10	9 41	**2 35**	6 00	**8 45**
7#20	....	**3#45**	**1 05**	9 45	Ar **Santa Cruz** ... Lv	....	7 00	9 30	**2 25**	5 50	**8 35**
7 30	....	**4 00**	....	9 55	Lv **Santa Cruz** ... Ar	....	....	9#15	**2 15**	5#35	....
	....	**HS**	....	HS	**Freedom Blvd, Turnout**	....	....				....
8 00	....	**4 30**	....	10 25	Ar **Watsonville** ... Lv	....	....	8 45	**1 46**	5 05	....
8 02	....	**4 32**	....	10 27	Lv **Watsonville** ... Ar	....	....	8 40	**1 43**	5 03	....
8 13	....	**4 43**	....	10 38	Moss Landing ...	....	....	8 26	**1 32**	4 52	....
8 19	....	**4 49**	....	10 44	Castroville ...	....	....	8 20	**1 26**	4 46	....
8 35	....	**5 05**	....	11 00	Ar **SALINAS, CAL** ... Lv	....	....	8 05	**1 10**	4 30	....
8 55	....	***6 00***	....	*11 00*	Lv *Salinas* ... Ar	....	....	*5 50*	***12 55***	4 10	....
11 50	....	***9 00***	....	***2 00***	Ar *San Luis Obispo* ... Lv	....	....	*3 15*	*9 50*	1 30	....
2 20	....	***11 40***	....	***4 45***	Ar *Santa Barbara* ... Lv	....	....	*12 50*	*6 40*	10 40	....
3 31	....	*12 50*	....	***5 50***	Ar *Oxnard* ... Lv	....	....	***11 35***	*5 10*	9 25	....
	....	↓	....	***6 55***	Ar *North Hollywood* ... Lv	....	....	***10 10***	↑	8 40	....
505	....	*2 20*	....	***7 35***	Ar *Los Angeles, Cal.* ... Lv	....	....	***9 30***	*3 45*	8 00	....

REFERENCE MARKS

All service daily except as noted.
LIght faced figured indicate A.M.
Dark faced figures indicate P.M.

Times shown in ITALICS indicate service via connecting schedule or schedules.
#—Rest stop ||—Meal stop

1. John lives in Sunnyvale and works in San Francisco. His work location is two blocks from the bus station. He has to be at work at 8 A.M. Which bus (schedule number) should he take to work? ______________

2. How long is the bus ride to work? ______________________________

3. He leaves work at 4:30 P.M. each day. Which bus (schedule number) should he take home? ______________

4. How long is the bus ride home? ______________________________

5. One Friday afternoon after work, John decided to take the bus to Santa Cruz to visit some friends. What time did he arrive in Santa Cruz? ______________

6. How long was the bus ride from San Francisco to Santa Cruz? ______________

(continued)

Name ______________________________ Date ______________

10. Travel by Bus *(continued)*

7. John spent the night in Santa Cruz. The next morning he took the 9:55 bus (schedule number 3491) to Los Angeles. What time did he arrive in Los Angeles? __________

8. Did he arrive in Los Angeles in the A.M. or P.M.? __________

9. How long was the bus ride to Los Angeles? ______________________________

10. John had to be back to Sunnyvale the next day no later than 5 P.M. for a date. Which bus (schedule number) did he have to take from Los Angeles? __________

11. How long was the bus ride back to Sunnyvale? ______________________________

12. How much time was John able to spend in Los Angeles? ______________________________

Name ______________________________ Date ______________

11. Travel by Air

Part I

When you travel by air, you will need to know about different time zones. In the first row of the box are four time zones and the name of a city in each zone. The second row shows that when it is 6 A.M. in Los Angeles, it is 7 A.M. in Denver, and so on. Fill in the blanks in the box.

Hint: When it is 8 A.M. in Los Angeles, it is one hour later in Denver.

PacificDaylightTime (PDT) (LosAngeles)	MountainDaylightTime (MDT) (Denver)	CentralDaylight Time (CDT) (Chicago)	EasternDaylightTime (EDT) (Philadelphia)
6 A.M.	7 A.M.	8 A.M.	9 A.M.
8 A.M			11 A.M.
11 A.M.	12 noon		

Part II

In the box is part of an airline schedule showing flights from Austin to three other cities. Read it and answer the questions.

Hint: Since Austin is on Central Daylight Time, all times shown for leaving are in Central Daylight Time. Arrival times are shown according to time zone of the city you are arriving in.

FROM AUSTIN (CDT)

To: **DENVER** (MDT)

Leave	Arrive	Flight	Operates	One-Way Fare Business Class	Coach
7 00a	9 05a	22/60	Daily	$175.00	$90.00
8 00a	10 05a	146/52	Daily		
10 00a	12 05p	126/68	Daily		
11 10a	1 05p	228/176	Daily		
1 05p	3 05p	8/170	Daily		

To: **MEMPHIS** (CDT)

Leave	Arrive	Flight	Operates	One-Way Fare Business Class	Coach
7 00a	9 30a	22/20	Daily	$169.00	$85.00
8 00a	10 25a	146/114	Daily		
11 10a	1 35p	228/204	Daily		
2 10p	4 35p	44/216	Daily		

To: **NEW YORK-NEWARK AIRPORT** (EDT)

Leave	Arrive	Flight	Operates	One-Way Fare Business Class	Coach
7 00a	12 40p	22/2	Daily	$390.00	$190.00
7 00a	1 00p	22/20	Daily		
8 30a	1 30a	18	Daily		
10 00a	3 20p	126/24	Daily		
10 00a	330p	126/4	Daily		
11 10a	4 30p	228/16	Daily		
11 10a	5 20p	228/28	Daily		

(continued)

Name ______________________ Date ______________

11. Travel by Air *(continued)*

1. If you had a meeting in Memphis at 4 P.M., what is the latest flight you could take from Austin?

   ____

2. If you had a meeting in New York at 3 P.M., what is the latest flight you could take from Austin?

   ____

3. If you took flight 8/170 from Austin to Denver, what time would it be in *Austin* when you arrived in Denver? ______________

4. If you took flight 126/4 from Austin to New York, what time would it be in *Austin* when you arrived in New York? ______________

5. How much would you save by going coach instead of business class from Austin to New York?

   $ ______________

6. How much would you save by going coach instead of business class from Austin to Denver?

   $ ______________

7. How much would you save by going coach instead of business class from Austin to Memphis?

   $ ______________

Name ______________________________ Date ______________

12. Motel Costs

In the box is part of a travel guide prepared by an automobile club. It describes two motels in a town. Read it and answer the questions.

Hint: The prices for one room for two people for one night at the Royal Motel are circled. You may be charged anywhere from $55.00 to $60.00 for one night. If you were asked how much it would cost two people for *two* nights, you would answer $110.00 to $120.00.

In-Season Rates One Room	One Person	Two Persons	Three Persons	Four Persons
Royal Motel	45–50	(55–60)	65—70	75–80

Two levels. Cheerful, well-kept, comfortable units: A/C; TV; free outside calls; outdoor pool; children under 12 free; no pets; all major credit cards accepted. 120 units. RATING: GOOD

- Restaurant; 24 hours; dinners $8.95–$15.95

	One Person	Two Persons	Three Persons	Four Persons
Hound Lodge	79–110	99–130	114–145	124–155

One level. Attractive, very well-kept one- and two-room units; A/C; cable TV; refrigerators in rooms; direct-dial phones; movie rentals; pets; indoor pool; 34 units. RATING: VERY GOOD

- Restaurant; dinner $12.00–$24.00

1. a. If you wanted a room for just yourself at the Royal Motel, how much would it cost you for one night? $ ______________

   b. How much would it cost you for three nights? $ ______________

2. a. If you wanted a room for yourself and your spouse at the Hound Lodge, how much would it cost for one night? $ ______________

   b. How much would it cost for two nights? $ ______________

3. a. If you wanted a room for yourself and your spouse as well as your children (aged 9 and 14) at the Royal Motel, how much would the total cost be for one night? $ ______________

   b. How much would it cost for four nights? $ ______________

(continued)

Name ______________________________ Date ______________

12. Motel Costs *(continued)*

4. How much more would it cost you for just yourself for one night at the motel with a "very good" rating than for a night at the motel with a "good" rating? $ ______________

5. If you got a room for yourself and your spouse, as well as a separate room for your 18-year-old child, at the Royal Motel, what would it cost you per night? $ ______________

6. a. If you ordered the most expensive dinner at the Hound Lodge, how much would it cost you?

   $ __________

   b. If you ordered the least expensive dinner at the Royal Motel, how much would it cost you?

   $ __________

7. If you, your spouse, and two children each ordered the most expensive dinner at the Royal Motel, what would be the total cost? $ __________

8. If three people traveling together each ordered the least expensive dinner at the Hound Lodge, what would be the total cost? $ __________

9. Which motel has more rooms? ______________________

10. If you were arriving late at night and wanted to eat, which motel has the restaurant that is more likely to be open? ______________________

Name ______________________________ Date ______________

13. Camping Costs

The Marciano family is spending part of their summer vacation camping at the Muskeegee Wilderness Campground. Use the following chart to help determine the costs associated with the Marcianos' vacation.

Type of Campsite	Cost (for 2 people)	Additional Costs
Hookup Services include electricity, running water, laundry service, rest rooms and shower facilities	$18.00 each night	$3.00 each night (per person)
Tent Additional services • Coin-op showers • Coin-op laundry (washer/dryer both included)	$12.00 each night $0.50 per 10 minutes $1.50 per load	$1.50 each night (per person) $0.25 per 5 minutes

1. a. What is the cost for Mr. and Mrs. Marciano to camp at the hookup campsite for three nights?

   $ ____________

   b. What is the cost for them to stay three nights at the tent campsite utilizing no additional services?

   $ ____________

2. a. How much would it cost Mr. and Mrs. Marciano each night to camp at the hookup site with their three children? $ ____________

   b. How much would it cost them to camp at the tent site each night with their three children?

   $ ____________

3. a. What are some benefits of camping at the hookup site? ______________________________

   b. What are some benefits of camping at the tent site? ______________________________

4. What is the cost to stay at the tent campsite one week if each of the five Marcianos takes a five-minute shower and doesn't wash laundry? $ ____________

5. If the Marcianos wash two loads of laundry and each person takes a daily five-minute shower, how much will they save by camping at the tent site instead of the hookup site for four nights? $ ____________

Name ______________________________ Date ______________

Section 2: Transportation and Vacation Costs Test

1. If gasoline to run a car with a small engine will cost you $468 per year and gasoline to run a car with a large engine will cost you $819 per year, how much will you save on gasoline per year by purchasing a car with a small engine? $ __________

2. The cash price for a used car is $4,788.00 and the deferred price is $5,445.66. How much will you save if you purchase the car in cash? $ __________

3. Marissa decides to take out a loan to pay for a new car whose total cost is $12,122.00. If her total monthly payments are $241.00, how many months will it take Marissa to pay off the loan after putting down a $2,000.00 deposit on the new car? ____________________

4. The price for a used motorcycle is $3,558.00. There is a 6% sales tax on all vehicles. How much is the tax on the motorcycle? $ __________

5. How much will it cost to lease a car for 36 months with a one-time down payment of $799.00 and regular payments of $186.00 per month? $ __________

6. A bus leaves Salinas at 6 P.M. and arrives in Los Angeles at 2:20 A.M. How long is the bus ride? ____________________

7. When it is 10 A.M. in New York (Eastern Standard Time), what time is it in Chicago (Central Standard Time)? __________

8. The cost to travel one way by air from Buffalo, N.Y., to Orlando, Fla., in business class is $326.00 and $160.00 in coach. How much would you save if you traveled round-trip in coach class? $ __________

9. At the ABC Motel, rooms for two people cost $74.00 per night. How much will it cost you and a friend if you share a room at the motel for three nights? $ __________

10. A campground rents space for $9.00 per night for the first person plus $1.75 for each additional person. How much will it cost for you and three friends to stay for four nights? $ __________

Section 3: Savings and Checking Accounts

TEACHER PAGES

14. Checking Account Deposit

Computational Skills
Addition, Subtraction

Mathematical Content
Currency

Procedure

1. Discuss with students the uses of a checking account and how it differs from a savings account. Be sure students understand the word "deposit."
2. In Part II, Rose Black will have to sign on the line under the words "Received cash returned from deposit" because she will be receiving cash from her deposit. She should wait to sign until she is with the bank teller.

Part I

UNITED BANK
Westlake Office
373 Oak Boulevard
Oaks, California 91360

90-3011
1221

Date 1/21 (current year here)
Received cash returned from deposit

Deposited for credit of Rose Black
22 First Street
Oaks, CA 91360

CASH	67.50
Checks by Bank No. 80-3113	123.84
Subtotal if Cash Returned from Deposit	
Less Cash Returned from Deposit	
TOTAL DEPOSIT	191.34

222‴301 2652041 02

Part II

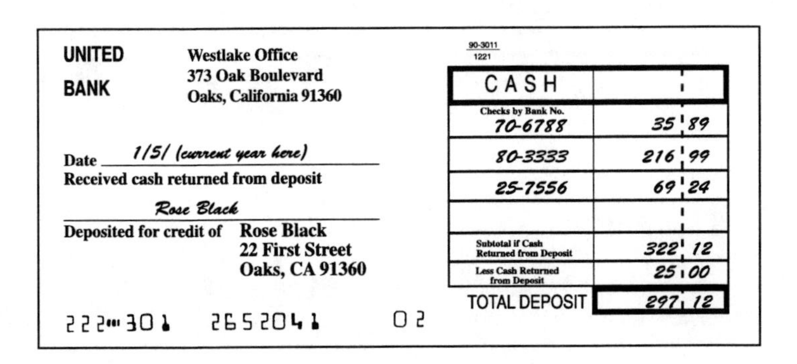

UNITED BANK
Westlake Office
373 Oak Boulevard
Oaks, California 91360

90-3011
1221

Date 1/51 (current year here)
Received cash returned from deposit
Rose Black

Deposited for credit of Rose Black
22 First Street
Oaks, CA 91360

CASH	
Checks by Bank No. 70-6788	35.89
80-3333	216.99
25-7556	69.24
Subtotal if Cash Returned from Deposit	322.12
Less Cash Returned from Deposit	25.00
TOTAL DEPOSIT	297.12

222‴301 2652041 02

15. Check Register

Computational Skills
Addition, Subtraction

Mathematical Content
Currency

Procedure

1. Point out to students that the automatic transfer (point 3) should be treated as a withdrawal. The amount should be listed under "Amount of check" and subtracted from the old balance to get the new balance.
2. Point out to students that the column with a check (✓) as a heading is used to check off canceled checks when they are received from the bank.

CHECK NO.	DATE	CHECKS ISSUED TO OR DESCRIPTION OF DEPOSIT	AMOUNT OF CHECK	√	AMOUNT OF DEPOSIT	PLEASE DEDUCT ANY PER CHECK CHARGES OR AUTOMATIC TRANSFERS	BALANCE FORWARD	
							254	25
185	1/27	To Bob Smith, M.D.	20.00			Check or Dep.	20	00
		For Office Visit—co-pay				Bal.	234	25
	1/28	To Deposit			1,220.52	Check or Dep.	1,220	52
		For Payroll Check				Bal.	1,454	77
186	1/29	To Holiday Hardware	63.59			Check or Dep.	63	59
		For Electric Drill				Bal.	1,391	18
187	1/30	To Safetimes Supermarket	67.89			Check or Dep.	67	89
		For Food				Bal.	1,323	29
	1/31	To Automatic Transfer	200.00			Check or Dep.	200	00
		For Savings				Bal.	1,123	29
	2/1	To Service Charge	3.00			Check or Dep.	3	00
		For				Bal.	1,120	29
	2/5	To Deposit			600.00	Check or Dep.	600	00
		For Cash				Bal.	1,720	29
188	2/6	To Frank's Place	39.99			Check or Dep.	39	99
		For Pants				Bal.	1,680	30
189	2/8	To Joe's Garage	567.60			Check or Dep.	567	60
		For Brake Repairs				Bal.	1,112	70

16. Checking Account Statement

Computational Skills

Addition, Subtraction

Mathematical Content

Tables/Graphs, Currency

Procedure

1. Be sure students understand the following terms before assigning the exercise: checking account statement, balance, deposit, return check charge, service charge, gap in check sequence, and withdrawals.
2. Students may need assistance with the entire exercise. A transparency of this exercise may be made for this purpose.
    1. $261.89 (See "Balance" under "Previous Statement.")
    2. $83.95 (See "New Balance.")
    3. $311.73 (See "Checking Account Daily Balances.")

4. $660.47 (See "Deposits" under "Totals for This Statement Period.")
5. $25.00 (See "Deposits.")
6. $5.30 (See "Withdrawals.")
7. $60.00 (See "Withdrawals"; note two withdrawals.)
8. $838.41 (Note that "Total for This Statement Period" includes "Returned Check Charge" and "Service Charge.")
9. $243.69 ($159.74 + $83.95 = $243.69)
10. $208.69 ($243.69 – $35.00 = $208.69)

17. Savings Account Book

Computational Skills
Addition

Mathematical Content
Tables/Graphs, Currency

Procedure

Discuss the concept of "interest" with the students. For example, banks pay interest to customers for saving with them; banks lend the money to others and charge them a higher rate of interest.

1. $1,751.04
2. $1,501.04
3. $1,001.75
4. $1,835.32 ($1,785.32 + $50.00 = $1,835.32)
5. $450.00 ($100.00 + $150.00 + $100.00 + $100.00 = $450.00)
6. $0
7. $540.00 ($420.00 + $120.00 = $540.00)
8. $1,835.32 ($1,785.32 + $50.00 = $1,835.32)
9. $990.00 ($450.00 for last year + $540.00 for January = $990.00)
10. $36.43 ($15.72 + $20.71 = $36.43)
11. $1,835.32

18. Costs of Bank Services

Computational Skills
Subtraction, Multiplication, Division

Mathematical Content
Percents/Proportions/Decimals, Currency

Procedure

1. Discuss direct deposit with students. Paychecks are automatically added to the balance of certain bank accounts. People can choose which accounts get deposits and how much of the paycheck goes into each account. Some banks give free or reduced fee services when checks are directly deposited.
2. Be sure students know how to multiply decimals before assigning this exercise. Review the example given in the Hint in Part II with students who may be deficient in this skill.

Part I

1. Plan A (Note: $.10 × 15 = $1.50 = monthly cost for checks under Plan A.)
2. Plan A (Note: $.09 × 25 = $2.25 = monthly cost for checks under Plan A.)
3. Plan B (Note: $.11 × 30 = $3.30 = monthly cost for checks under Plan A.)
4. $3.05 ($200.00 – $139.00 = $61.00 × .05 = $3.05)
5. $9.35 ($200.00 – $13.00 = $187.00 × .05 = $9.35)

Part II

6. a. $2.50 ($250.00 × .01 = $2.50)
   b. 25 (250 ÷ 10 = 25)
7. a. $3.00 ($300.00 × .01 = $3.00)
   b. 15 (300 ÷ 20 = 15)
8. a. $6.00 ($600.00 × .01 = $6.00)
   b. 12 ($600 ÷ 50 = 12)
9. a. $10.00 ($1,000.00 × .01 = $10.00)
   b. 10 (1,000 ÷ 100 = 10)

Section 3 Test

1. $792.45 ($852.45 – $60 = $792.45)
2. $429.98 ($129.34 + $345.66 = $475.00 – $45.02 = $429.98)
3. $424.77 ($167.23 + $192.54 + $65.00 = $424.77
4. $1,101.03 ($80 × 3 = $240; $991.91 – $240 = $751.91 + $352.62 = $1,104.53 – $3.50 = $1,101.03)
5. $487.50 ($650.00 – $97.50 – $65.00 = $487.50)
6. $309 ($300 × .03 = $9 + $300 = $309)
7. $415.79 ($378.63 + $35.00 + $2.16 = $487.50)
8. $381.19 ($590.00 + $8.94 – $217.75 = $381.19)
9. $1.53 (17 × $0.09 = $1.53)
10. $1.14 ($250 – $212 = $38 × .03 = $1.14)

Name ______________________ Date ______________

14. Checking Account Deposit

Part I

On January 2 of this year, Rose deposited $67.50 in cash and a check for $123.84 to her checking account. The bank number on the check she deposited was 80-3113. Fill out the checking account deposit slip for Rose.

Hint: Write the amount of cash to the right of the word "Cash." The dashed line should be used to separate dollars and cents.

UNITED BANK

Westlake Office
373 Oak Boulevard
Oaks, California 91360

Date ______________________
Received cash returned from deposit

____
Deposited for credit of

Rose Black
22 First Street
Oaks, CA 91360

222⑈301⑆ 2652041⑆ 02

90-3011
1221

Cash	
Checks by Bank No.	
Subtotal if cash returned from deposit	
Less cash returned from deposit	
TOTAL DEPOSIT	

Part II

On January 5 of this year, Rose went to the bank with a check for $35.89 (bank number 70-6788), a check for $216.99 (bank number 80-3333), and a check for $69.24 (bank number 25-7556). She wanted to deposit all but $25.00, which she wanted to receive in cash. Fill out the checking account deposit slip for Rose.

Hint: Add up the amount of the checks and write the sum to the right of "Subtotal." Write the amount Rose wants in cash to the right of "Less cash returned from deposit." The total deposit is the subtotal minus the amount of cash received.

UNITED BANK

Westlake Office
373 Oak Boulevard
Oaks, California 91360

Date ______________________
Received cash returned from deposit

____
Deposited for credit of

Rose Black
22 First Street
Oaks, CA 91360

222⑈301⑆ 2652041⑆ 02

90-3011
1221

Cash	
Checks by Bank No.	
Subtotal if cash returned from deposit	
Less cash returned from deposit	
TOTAL DEPOSIT	

Name ______________________________ Date ______________

15. Check Register

A check register is used to record all deposits into and withdrawals from a checking account. Part of Rick Lamonte's check register is shown in the box. Fill in the rest of it for him using this information:

1. On January 29, Rick wrote a check for $63.59 to Holiday Hardware for an electric drill.
2. On January 30, Rick wrote a check for $67.89 to Safetimes Supermarket for food.
3. On the last day of each month, the bank automatically transfers $200.00 from Rick's checking account to his savings account.
4. A service charge of $3.00 per month is automatically deducted from Rick's checking account. Rick deducts this amount in his register on the first day of each month.
5. On February 5, Rick transferred $600.00 from his savings account to his checking account.
6. On February 6, Rick wrote a check for $39.99 to Frank's Place for a pair for pants.
7. On February 8, Rick wrote a check for $567.60 to Joe's Garage for brake repairs on his car.

PLEASE DEDUCT ANY PER CHECK CHARGES OR AUTOMATIC TRANSFERS BALANCE FORWARD

CHECK NO.	DATE	CHECKS ISSUED TO OR DESCRIPTION OF DEPOSIT	AMOUNT OF CHECK	√	AMOUNT OF DEPOSIT			
							254	25
185	1/27	To Bob Smith, M.D.	20.00			Check or Dep.	20	00
		For Office Visit—co-pay				Bal.	234	25
	1/28	To Deposit			1,220.52	Check or Dep.	1,220	52
		For Payroll Check				Bal.	1,454	77
		To				Check or Dep.		
		For				Bal.		
		To				Check or Dep.		
		For				Bal.		
		To				Check or Dep.		
		For				Bal.		
		To				Check or Dep.		
		For				Bal.		
		To				Check or Dep.		
		For				Bal.		
		To				Check or Dep.		
		For				Bal.		
		To				Check or Dep.		
		For				Bal.		

Name ______________________________________ Date ____________________

16. Checking Account Statement

Below is Wayne Oppenheimer's checking account statement. It shows all the changes in Wayne's account. Read the statement and answer the questions.

INDEPENDENCE BANK

CHECKING ACCOUNT NUMBER:
086 05439 8924

STATEMENT PERIOD FROM **JUN 18–JUL 18**
PAGE **1 of 1**

Wayne C. Oppenheimer
3456 Park Street
Center, Iowa 52360

PREVIOUS STATEMENT		TOTAL FOR THIS STATEMENT PERIOD				NEW BALANCE
		CHECKS		DEPOSITS		
DATE	BALANCE	NO.	AMOUNT	NO.	AMOUNT	
06:17	**261:89**	**14**	**838:41**	**3**	**660:47**	**83.95**

Reference or Check No.	DATE	AMOUNT	Reference or Check No.	DATE	AMOUNT	Reference or Check No.	DATE	AMOUNT
DEPOSIT			**WITHDRAWAL**					
			250	**06:20**	**231:83**	**261**	**07:05**	**21:23**
			255*	**06:30**	**316:75**	**263***	**07:15**	**50:00**
43	**06:23**	**520:00**	**256**	**06:20**	**10:00**	**264**	**07:18**	**8:54**
44	**06:29**	**25:00**	**257**	**06:24**	**5:30**	**266***	**07:06**	**120:00**
45	**07:05**	**115:47**	**258**	**06:29**	**19:16**	**267**	**07:18**	**16:24**
			258	**06:27**	**6:36**	**268**	**07:15**	**10:00**
						R/C	**07:16**	**20:00**
			260	**06:27**	**10:00**	**S/C**	**07:18**	**3:00**

CHECKING ACCOUNT DAILY BALANCES 06–20 THRU 07–18

DATE	BALANCE	DATE	BALANCE
06:20	**30:06**	**07:05**	**311:73**
06:23	**550:06**	**07:06**	**191:73**
06:24	**544:76**	**07:11**	**171:73**
06:27	**528:40**	**07:15**	**111:73**
06:29	**534:24**	**07:18**	**83:95**
06:30	**217:49**		

R/C Returned Check Charge **S/C Service Charge** *** Gap in Check Sequence**

(continued)

Name ______________________________ Date ______________

16. Checking Account Statement *(continued)*

1. How much did Wayne have in his checking account on June 17? $__________
2. How much did Wayne have in his checking account on July 18? $__________
3. How much did Wayne have in his checking account on July 5? $__________
4. How much did Wayne put in his checking account between June 18 and July 18? $__________
5. How much did Wayne put in his checking account on June 29? $__________
6. How much was taken out of Wayne's account on June 24? $__________
7. How much was taken out of Wayne's account on July 15? $__________
8. What was the total amount taken out of Wayne's account between June 18 and July 18?

   $__________
9. On July 19, Wayne deposited $159.74. What was his new balance? $__________
10. On July 20, Wayne withdrew $35.00. What was his new balance? $ __________

Name ______________________________ Date ______________

17. Savings Account Book

On August 11 of last year, Judy opened a savings account with a deposit of $1,785.32. The first page in her bank book is shown in the box. It shows everything that happened in her account through January of this year.

Hints: The withdrawal column shows amounts Judy took out of her savings. Interest is money the bank paid Judy for saving in the bank. Additions are amounts Judy put into her account. Balance is the amount Judy had in her account after a withdrawal was made, interest was paid, or an addition was made.

	DATE	WITHDRAWAL	INTEREST	ADDITIONS	BALANCE
1	AUG 11			1,785.32	1,785.32
2	SEPT 11			50.00	1,835.32
3	SEPT 24	100.00			1,735.82
4	SEPT 27		15.72		1,751.04
5	OCT 29	150.00			1,601.04
6	NOV 24	100.00			1,501.04
7	NOV 26	100.00			1,401.04
8	DEC 27		20.71		1,421.75
9	JAN 17	420.00			1,001.75
10	JAN 22	120.00			881.75

1. After interest was added on September 27, how much did Judy have in her account? $____________
2. After Judy make a withdrawal on November 24, how much did she have in her account? $____________
3. After Judy made a withdrawal on January 17, how much did she have in her account? $____________
4. How much did Judy deposit into her account during last year? $____________
5. How much did Judy withdraw from her account last year? $____________
6. How much did Judy deposit into her account during January of this year? $____________
7. How much did Judy withdraw from her account during January of this year? $____________
8. How much did Judy deposit into her account for the entire period? $____________
9. How much did Judy withdraw from her account for the entire period? $____________
10. How much interest was credited to Judy's account for the entire period? $____________
11. What was the highest balance Judy had? $____________

Name ______________________________ Date ______________

18. Costs of Bank Services

Part I

Answer these questions about checking account plans.

Hint: Some banks offer checking account plans in which you are charged a certain amount for each check you write. If a bank charges 7¢ per check and you write about 20 checks per month, the checking account will cost you about $1.40 per month ($.07 x 20 = $1.40).

1. Tom's bank offers two different checking plans. In Plan A, Tom would have to pay 10¢ for every check he writes. In Plan B, he would pay $2.00 per month regardless of how many checks he writes. Tom writes about 15 checks a month. Which plan should Tom choose?

   ____

2. Julia's bank offers two different checking plans. In Plan A, Julia would have to pay 9¢ for every check she writes. In Plan B, she would pay $3.00 per month regardless of how many checks she writes. She writes about 30 checks a month. Which plan should Julia choose?

   ____

3. Maria's bank offers two different plans. In Plan A, Maria would have to pay 11¢ for every check she writes. In Plan B, she would pay $2.50 per month regardless of how many checks she writes. She writes about 30 checks per month. Which plan should Maria choose?

   ____

4. Damien's bank offers free checking when paychecks are directly deposited and a minimum balance of $200.00 stays in the checking account. However, a 5% surcharge is deducted for every dollar drawn under the $200.00. If Damien had direct deposit at his job but lets his minimum balance fall to $139.00, what will his bank charge him?

   $______________________________

5. Jeannine has the same checking account as Damien. This past month, she withdrew all but $13.00 from her checking account. How much was she charged?

   $______________________________

(continued)

Name ______________________________ Date ______________

18. Cost of Bank Services *(continued)*

Part II

Answer these questions about traveler's checks. Assume that traveler's checks cost 1% of the amount of the checks.

Hint: To find 1% of an amount, multiply the amount by .01. For example, 1% of $200.00 equals $2.00 since

```
 $200.00
   x .01
 $2.0000
```

6. a. Sampson bought $250.00 worth of traveler's checks. How much did the checks cost him?

   $____________

   b. Sampson asked for $250.00 worth of 10-dollar checks. How many checks should he have received?

   ____

7. a. Alice bought $300.00 worth of traveler's checks. How much did the checks cost her?

   $____________

   b. Alice asked for $300.00 worth of 20-dollar checks. How many checks should she have received?

   ____

8. a. David bought $600.00 worth of traveler's checks. How much did the checks cost him?

   $____________

   b. David asked for $600.00 worth of 50-dollar checks. How many checks should he have received?

   ____

9. a. Sarah bought $1,000 worth of traveler's checks. How much did the checks cost her?

   $____________

   b. Sarah asked for $1,000 worth of 100-dollar checks. How many checks should she have received?

   ____

Name ____ Date ______________________

Section 3: Savings and Checking Accounts Test

1. Alicia's paycheck is $852.45. She wants to deposit all of it into her checking account except for $60.00. How much will she be depositing? $ __________

2. Larry's checking account balance was $129.34 before he deposited $345.66 and wrote a check for $45.02. What is his new balance? $ __________

3. During the month of July, Sandy made three deposits in her checking account. The amounts were $167.23, $192.54, and $65.00. What was the total amount she deposited during July? $ __________

4. In the beginning of the month, Mr. Stockholm's checking account balance was $991.91. During the month, he made three withdrawals of $80.00 each, deposited checks for $21.50, $327.00, and $4.12, and was charged a $3.50 service fee. What is his new balance now? $ __________

5. Deborah has her $650.00 weekly paycheck directly deposited into her checking account. Her bank automatically transfers 15% of her paychecks to her savings account and another 10% is deducted as a contribution to her favorite charity. What is the total amount of Deborah's paycheck that is deposited into her checking account? $ __________

6. Kendra opened a savings account with $300.00, which she received as a birthday gift from various relatives. The savings account pays 3% interest each month. What is Kendra's balance at the end of the first month? $ __________

7. Pat's savings account balance was $378.63 before he deposited $35.00 and was credited with $2.16 in interest. What is his new balance? $ __________

8. During May, Ms. Tedesco added $590.00 to her savings account, was credited with $8.94 interest, and withdrew a total of $217.75. What is the total change in Ms. Tedesco's savings account balance for the month of May? $ __________

9. Fran's bank charges her $.09 for each check she writes. During December, she wrote 17 checks. How much did the bank charge her during December? $ __________

10. A certain bank offers free checking with a minimum balance of $250.00 and charges 3% for each dollar that falls below that minimum balance. If Lana's balance falls to $212.00, how much will her bank charge as a penalty? $ __________

Section 4: Household Budgeting

19. Housing Costs I

Computational Skills
Subtraction, Multiplication

Mathematical Content
Percents/Proportions/Decimals, Currency

Procedure

1. You may want to discuss the pros and cons of renting versus owning a house or apartment or condo.
2. Before assigning Part II, review multiplication of a dollar amount by a decimal with students. Question 7 in Part II may be used for this purpose.

Part I

1. Apartment A
2. $895
3. $585
4. $6,780
5. a. $145
   b. $1,740
6. Answers will vary. Acceptable answers include better location, larger rooms, more appliances provided.

Part II

7. House F ($4,250.00 × .25 = $1,062.50)
8. House G ($5,130 × .25 – $1,282.50)
9. House H ($6,900 × .25 = $1,725.00)
10. No ($1,100.00 × .25 = $275.00)
11. No ($1,880 × .25 = $470)

20. Housing Costs II

Computational Skills
Addition, Multiplication, Division

Mathematical Content
Currency

Procedure

1. Discuss these terms in Part I with students before assigning the exercise: “payable prior to occupancy,” “security deposit,” “key deposit,” and “cleaning fee.”
2. Discuss these terms in Part II with students before assigning the exercise: “mortgage,” “real estate taxes,” and “fire insurance.”

3. Discuss factors which determine the amount of mortgages (amount of loan, interest rate, fixed or adjustable rate, terms of loan, etc.); fire insurance or home owner's insurance (does home have smoke alarms, what is house made of, have fireplace, woodstove etc.); real estate tax (for example, mill rate set by town, assessed value of house).
4. Instruct students to carry division problems to three decimal places and round to two.

Part I

1. \$855.00
2. \$250.00 (\$200.00 + \$50.00 = \$250.00)
3. \$1,715.00 (\$855.00 + \$855.00 + \$5.00 = \$1,715.00)
4. None (Note: He paid it prior to occupancy.)
5. \$10,260.00 (\$855.00 × 12 = \$10,260.00)

Part II

6. a. \$57.69 (\$692.25 ÷ 12 = \$57.69)
   b. \$160.78 (\$1,929.33 ÷ 12 = \$160.78)
   c. \$1,649.37 (\$1,430.90 + \$57.69 + \$160.78 = \$1,649.37)

21. Savings for Future Expenses

Computational Skills	**Mathematical Content**
Addition, Multiplication, Division	Currency

Procedure

1. Go over items, 1, 5, and 7 with less advanced students before assigning the exercise.
2. Remind students to compute their answers to three decimal places and round off to two.
3. Have students list other expenses they may encounter later on in their lives. Discuss the importance of saving money now for use later.

1. \$104.73 (\$1,256.78 ÷ 12 = \$104.73)
2. \$28.80 (\$245.60 + \$100.00 = \$345.60; \$345.60 ÷ 12 = \$28.80)
3. January 1 of the following year (\$250 ÷ 25 = 10 months to save)
4. \$54.17 (2 × 6 = 12 paydays; \$650.00 ÷ 12 = \$54.17)
5. \$62.50 (2 × 8 = 16 paydays; \$1,000.00 ÷ 16 = \$62.50)
6. Answers will vary.

22. Budget Planning

Computational Skills
Addition, Subtraction

Mathematical Content
Currency

Procedure

1. The first four items in the budget are the most important. Students can estimate rent by consulting the classified ads in a local newspaper; they can obtain estimates of utilities by talking with representatives of utility companies or talking with single people who live in apartments; parents can help students estimate the monthly costs of food and beverages for a single person; students can estimate transportation by considering the cost of local bus service.
2. Discuss with students the relative importance of the remaining categories. Point out to students that many experts suggest saving the equivalent of six months' salary as soon as possible.

   Answers will vary.

Section 4 Test

1. $8,100 ($675 × 12 = $8,100)
2. $496.75 ($1,987 × .25 = $496.75)
3. $1,480 ($665 × 2 = $1,330 + $150 = $1,480)
4. $1,137.50 ($1,111 + $779 = $1,890 ÷ 12 = $1,57.50 + $980 = $1,137.50)
5. No ($1,500/mo. = $18,000/yr.; 25% × $50,000 = $12,500; $12,500 < $18,000)
6. $239.40 ($19.95 × 12 = $239.40)
7. $59.09 ($650 ÷ 11 = $59.09)
8. Almost 30 months ($14,916 ÷ $500 = 29.8)
9. $398.50 ($1,788.50 – $1,390 = $398.50)
10. $6,384 ($900 × 12 = $10,800; $23,568 – $10,800 = $12,768 ÷ 2 = $6,384)

Name ______________________ Date ______________

19. Housing Costs I

Part I

In the boxes are four advertisements offering apartments for rent. Each ad shows the amount of rent per month. Answer the questions.

A.

C.

B.

D.

1. Which apartment is the least expensive per month? ______________
2. How much is the rent per month for Apartment C? $ ______________
3. How much more per month is the rent for Apartment D than for Apartment A? $ ______________
4. How much is the rent per year for Apartment A? $ ______________
5. a. How much more per month is the rent for Apartment C than Apartment B? $ ______________

   b. How much more per year is the rent for Apartment C than Apartment B? $ ______________
6. Name two possible reasons why the rent is higher for the two-bedroom apartment than it is for the three-bedroom apartment. ____

   ____

(continued)

Name ______________________________ Date ______________

19. Housing Costs I *(continued)*

Part II

In the boxes are four advertisements offering houses for rent. Most people can afford to spend about 25% of what they earn for housing. Use the 25% rule when answering the following questions.

Hint: To find 25% of an amount, multiply by .25.

E.

$700 + util. Redec., 2br., carport, stv. / ref. w/w carpt., tot ok, Bkr fee. 213/753-5461

G.

$1,200 Eagle Rock area, 2 br., 2½ baths, frplc, air, huge yard, 1 yr lease, 2 car garage, cent. air 213/497-9798

F.

$975 util incl'd, 2br., 2 bath. Kids OK, crpts, drps, yard. HOME RENTALS FEE 213/380-8796

H.

$1,725 No fee. 4-br 2-ba. Fenced, pool, brkfst bay, good neighborhood. 714/780-7898

7. Mr. and Mrs. Goodman together earn $4,250.00 per month. Which is the most expensive house they can afford to rent? ______________
8. Mr. Wright earns $5,130 per month. Which is the most expensive house he can afford to rent?

   ____
9. Ms. Encinas earns $6,900 per month. Which is the most expensive house she can afford to rent?

   ____
10. Mr. Lang earns $1,100 per month. Can he afford to rent House E? ______________
11. Miss Steinman earns $1,880.00 per month. Can she afford to rent House E? ______________

Name ____ Date ____________________

20. Housing Costs II

Part I

In the box is part of a rental agreement that Lee Cheng signed when he rented an apartment. Read it and answer the questions.

	RECEIVED	Payable Prior to Occupancy
Rent for the period from *May 1* to *May 31*	$	$ *855.00*
Last month's rent	$	$ *855.00*
Security deposit	$ *200.00*	$
Key deposit	$	$ *5.00*
Cleaning charge	$ *50.00*	$
Other *None*	$	$

1. How much rent does Lee pay each month? $____________

2. How much money did the landlord receive at the time the agreement was signed? $____________

3. How much more will Lee have to give the landlord before moving in? $____________

4. When Lee decides to move, how much money will he have to pay the landlord at the beginning of his last month in the apartment? $____________

5. How much does Lee pay per year for the apartment? $____________

Part II

Answer the following questions about costs of buying a house.

6. a. Jill pays $1,430.90 per month to pay off the mortgage (loan) on the house she is buying. She must also buy fire insurance at $692.25 per year and real estate taxes at $1,929.33 per year. How much should she save each month in order to pay the fire insurance bill, which is sent only once a year?

   $____________

   b. How much should Jill save each month in order to pay the real estate tax bill, which is sent only once a year?

   $____________

   c. How much is Jill's total monthly housing cost including mortgage, fire insurance, and real estate taxes?

   $____________

Name ______________________________ Date ______________

21. Saving for Future Expenses

Answer the following questions.

1. Rena pays $1,256.78 per year for automobile insurance. She is paid monthly. How much should she save each month for her insurance?

   $__________

2. Trevor's parents live in another city. He would like to visit them in 12 weeks. The plane ticket will cost $245.60, and he will need another $100.00 for other expenses on the trip. He is paid weekly. How much should he save each week for the trip?

   $__________

3. Mr. Nuono would like to buy a motorcycle. He will need $250 for a down payment. If he saves $25 per month beginning on April 1, when will he have the $250? (Hint: Divide the amount he needs by the amount he will save per month to find out how many months he will need to save.)

   ____

4. Georgia would like to buy a new sofa for $650.00 in six months. She is paid on the first and fifteenth of each month. How much should she save each payday for the sofa? (Hint: She has 12 paydays to save for the sofa.)

   $__________

5. Bryan is getting married in eight months. He will need $1,000 for the honeymoon. He is paid on the first and fifteenth of each month. How much should he save each payday for the honeymoon?

   $__________

6. Think of three items you would like to buy that cost more than $25.00 each. List them in the first column. In the second column, write how much they will cost. The third column tells you how much to assume you can save per month. Figure out how many months it will take you to save enough money for each, and write your answers in the last column.

Items you would like to buy	Cost	Amount you can save per month	Numbers of months you will have to save
A. __________	$ ______	$ 8.00	__________
B. __________	$ ______	$12.00	__________
C. __________	$ ______	$20.00	__________

Name ______________________________ Date ______________

22. Budget Planning

Jenna recently graduated from high school and was hired by the ABC Company. After taxes and other deductions, she brings home $965.00 per month. She is living with her parents but wants to rent a furnished apartment. Assume that she has already moved into an apartment in your area.

Find out what the different items cost in your area. Keep in mind that Jenna does not own a car but takes a bus to work. She would like to buy a car as soon as possible. Eventually she would like to buy her own furniture. The ABC Company pays for her health insurance.

MONTHLY BUDGET	
	Amount per month
Total Income	$__________
Rent..	$__________
Utilities..	__________
(Telephone with Caller ID $________, Gas $________, Electric $ ________, Cable $ __________, Internet access $__________)	
Food and beverage	__________
Home owner's insurance	__________
Transportation	__________
Credit card bills	__________
Clothing	__________
Recreation / vacation	__________
Gifts / contributions	__________
Future goals / savings	__________
Emergencies / savings	__________
Total	$__________

Name ______________________________ Date ______________

Section 4: Household Budgeting Test

1. How much of Taylor's $39,935 annual salary is put toward his rental expense of $675.00 per month? $__________
2. Ernesto earns $1,987.00 per month. He wants to spend about 25% of his earnings on housing. How much is 25% of his monthly earnings? $__________
3. Lucy wants to rent an apartment for $665.00 per month. When she moves in, she will have to give the landlord the first month's rent, the last month's rent, and a security deposit of $150.00. What is the total amount she will have to give the landlord at move-in? $__________
4. Miss Rosco pays $980.00 per month toward her mortgage. She also pays $779.00 per year for flood insurance and $1,111.00 per year on real estate taxes. If she saves an equal amount each month to pay these housing costs, what total would she need to save? $__________
5. If Mr. and Mrs. Kreuger earn a combined yearly salary of $50,000, can they afford to spend 25% of their income toward a house that rents for $1,500.00 per month? Why or why not?
   ____
6. Bill pays $19.95 each month for Internet access. What is his yearly charge for this service? $__________
7. Rose is planning a vacation that will cost $650.00. She has 11 months to save for the vacation. How much should she save each month? $__________
8. Teresa is saving $500.00 each month for a new car. If the car will cost $14,916.00 after license, tax, and registration fees, how long will it take Teresa to accumulate enough to pay for the car in full? __________
9. Jake earns $1,788.50 per month after taxes. He spends $1,390.00 per month on rent, utilities, food, and transportation. How much does he have left after paying for these expenses? $__________
10. Stephanie Thisson's take-home pay for the previous year was $23,568.00. Last year, her monthly expenses averaged $900.00. After deducting expenses, she put half of her remaining earnings toward mutual fund investments. How much of her take-home pay was invested? $__________

Section 5: Buying on Credit

23. Credit Union Loan

Computational Skills
Subtraction, Multiplication

Mathematical Content
Tables/Graphs, Currency

Procedure

Before assigning the exercise, be sure students understand what a credit union is and understand the terms defined under the heading *Terms to Understand.*

1. \$70.62
2. \$1,694.88 (\$70.62 × 24 = \$1,694.88)
3. Answers will vary. They should indicate that the credit union charges a finance charge in order to pay the costs of running the credit union and to pay interest to the employees who save their money at the credit union. The money the credit union lends, for the most part, is money that employees save there.
4. \$50.40
5. \$14.21
6. Because his balance was less in September than in August
7. \$162.65 (\$1,500.00 – \$1,337.35 = \$162.65) Note: Students may be interested to know that if Mr. Rollings borrowed an additional \$162.65 at this point, he could have the original loan refinanced. That is, technically he would borrow another \$1,500.00, use \$1,337.35 to pay off the first loan, and have \$162.65 left for himself. In practice, the original loan would be paid off by the credit union staff so that he would never actually receive "in hand" the additional \$1,500.00. In this way, he would still have only one loan payment of \$70.62 per month.

24. Credit Card Buying

Computational Skills
Addition, Multiplication

Mathematical Content
Tables/Graphs, Percents/Proportions/Decimals, Currency

Procedure

1. Note that in Part I, students are always to round up to the next whole dollar. In Part II, they should compute their answers to three decimal places and round off to two.
2. Discuss these terms with students: balance (amount remaining—for credit card buying, this is almost always the amount owed); minimum payment (the smallest amount that may be paid); finance charge (cost of borrowing or interest); late charge fee; credit line (limit).

Part I

1. $12.00
2. $9.50
3. $15.00
4. $32.00 ($795.65 × .04 = $31.83 = $32.00)
5. $44.00 ($995.65 + $95.68 = $1,091.33; .04 × $1,091.33 = $43.65 = $44.00)

Part II

6. $5.10 ($340 × .015 = $5.10)
7. $7.88 ($525.33 × .015 = $7.879 = $7.88)
8. $10.95 ($729.76 × .015 = $10.946 = $10.95)
9. $17.52 ($1,000.00 × .015 = $15.00; $252.32 × .01 = $2.523 = $2.52; $15.00 + $2.52 = $17.52) Note: Only 1% is charged on amounts in excess of $1,000.00.

25. Home Mortgage

Computational Skills

Subtraction, Multiplication

Mathematical Content

Tables/Graphs, Currency

Procedure

1. Point out that a bank charges interest on a loan but pays interest on savings. In this exercise, interest is what the bank is charging Mr. Chavez for the loan.
2. Point out that a large amount of money is being paid solely toward a finance charge. Discuss the option of refinancing at a better interest rate in the future.

1. $314.53
2. $37,752.87
3. $37,729.35
4. $23.52
5. $291.01
6. Because it will be harder for him to obtain other loans in the future if he makes late payments on this one
7. $113,230.80 (30 × 12 = 360 payments; 360 × $314.53 = $113,230.80)
8. $75,230.80 ($113,230.80 – $38,000 = $75,230.80)

26. Car Loan

Computational Skills
Multiplication, Division

Mathematical Content
Tables/Graphs, Whole Numbers, Currency

Procedure

1. Before assigning this exercise, go over the material in the box with students. Have them draw up a list of terms they do not understand and provide them with definitions. Some of the terms they may not understand are:

vehicle price—cost of car, motorcycle, etc.

down payment—amount paid by Miss Post before she took out the loan; in this case, Miss Post traded in her old car and used this as her down payment

less lien—minus the amount she still owes on her trade-in; in this case, she does not owe anything

trade-in (net agreed value)—worth of the trade-in after subtracting the amount still owed

unpaid balance of cash price—amount owed on new car after subtracting the down payment

incl.—included; in this case, included in the cost of the new car

unpaid balance—amount owed on new car after adding in costs of insurance and official fees

deferred payment—amount to be paid later but not as part of monthly payments. For example, the dealer and bank may let you pay only part of your down payment when you buy the car and the rest several weeks or months later

amount financed—amount borrowed to make up the difference between the down payment and unpaid balance

finance charge—cost of borrowing money to buy the car

annual percentage rate—rate of interest; determines amount of finance charge

total of payments—total amount Miss Post will pay, not including her down payment

48 installments—48 payments

successive monthly payments—payments made each month

deferred payment price—total of payment and down payment

2. Students of varying abilities may need help with the entire exercise.
    1. \$5,600.00
    2. \$5,936.00
    3. \$1,950.00
    4. None
    5. \$3,986.00
    6. \$4,096.00
    7. \$1,474.40
    8. 4 years (48 ÷ 12 = 4)
    9. \$116.05
    10. \$1,392.60 (12 × \$116.05 = \$1,392.60)
    11. \$7,520.40
    12. Answers will vary.

Section 5 Test

1. \$229.32 (\$88.33 × 4 = \$353.32; \$1,000 – \$876 = \$124; \$353.32 – \$124 = \$229.32)
2. \$863.42 (\$53.5 – \$40.33 = \$13.17; \$876.59 – \$13.17 = \$863.42)
3. \$5.40 (\$359.99 × .015 = \$5.40)
4. \$31.01 (\$775.35 × .04 = \$31.01)
5. \$15.66 (\$1,000 × .015 = \$15; \$660 × .01 = \$0.66; \$15 + \$0.66 = \$15.66)
6. \$12,545 (\$93,000 – \$85,553 = \$4,747; \$17,292 – \$4,747 = \$12,545)
7. \$315,597.60 (\$876.66 × 30 × 12 = \$315,597.60)
8. \$9,898 (\$14,113 – \$4,215 = \$9,898)
9. \$5,206 (\$7,600 – \$2,394 = \$5,206)
10. \$15,264 (\$299 × 3 × 12 = \$10,764 + \$4,500 = \$15,264)

Name ______________________________ Date ______________

23. Credit Union Loan

Mr. Rollins may borrow up to $1,500.00 anytime he needs it from his credit union at work. In July, he borrowed the full amount. He agreed to pay it back by letting the credit union deduct a certain amount from his paycheck each month for two years. In the box is part of a statement Mr. Rollins received from the credit union. Read it and answer the questions.

Terms to Understand: transaction (exchange of money); finance charge (money Mr. Rollins must pay to the credit union for giving him the loan); principal (money that goes toward paying off the loan); balance (amount Mr. Rollins owes the credit union); payroll deduc. (payroll deduction—amount taken out of Mr. Rollins's paycheck and given to the credit union).

DATE			TRANSACTION	AMOUNT	FINANCE CHARGE	PRINCIPAL	BALANCE
MO	DAY	YR					
07	01	8	LOAN				1500.00
07	15	8	LOAN PAYMENT—PAYROLL DECUC—	70.62	20.22	50.40	1449.60
08	15	8	LOAN PAYMENT—PAYROLL DECUC—	70.62	14.78	55.84	1393.76
09	15	8	LOAN PAYMENT—PAYROLL DECUC—	70.62	14.21	56.41	1337.35
09	30	8				LOAN BALANCE	1337.35

1. How much is deducted from Mr. Rollins's paycheck each month? $__________
2. What is the total amount that will be deducted during the two years of the loan? $__________ (Hint: Twenty-four deductions will be made during the two years.)
3. Why is the answer to question 2 more than the $1,500.00 that Mr. Rollins borrowed?

____

____

4. How much of the July deduction went to pay off the amount Mr. Rollins borrowed? $__________
5. How much of the September deduction was a finance charge? $__________
6. Why was the amount of the finance charge for September less than the amount for August?

____

____

7. If Mr. Rollins wanted to take out another loan on September 30, how much could he borrow?

   $__________

Name ______________________________ Date ______________

24. Credit Card Buying

Part I

When you buy items with a credit card, you will be billed. When you are billed, you usually will be given a choice between paying the full amount or just making a payment. The material in the box tells how much the payments will be if you use one credit card. Read it and answer the questions.

Hints: To find 4% of an amount, multiply by .04. Since "all payments are adjusted to the next dollar amount," all of your answers in this part will end with two zeros.

Balance	Minimum Monthly Payment	Balance	Minimum Monthly Payment
$210.00	$10.00	$410.00	$15.00
$250.00	$11.00	$440.00	$16.00
$290.00	$12.00	$470.00	$17.00
$340.00	$13.00	$504.00	$18.00
$380.00	$14.00		

Over $504, minimum payment will be 4% of the highest account balance. All payments are adjusted to the next dollar amount. Minimum payment $10.00.

1. Tom charged a stereo that cost $265.95 This was all he owed on the credit card. How much will the minimum monthly payment be the first month? $____________
2. Laura charged a blouse that cost $9.50. This was all she owed on the credit card. How much will the monthly payment be the first month? $____________
3. Julian charged a color TV that cost $381.20. This was all he owed on the credit card. How much will the minimum monthly payment be the first month? $____________
4. Minnie charged living room furniture that cost $795.65. This was all she owed on the credit card. How much will the minimum monthly payment be the first month? $____________
5. Philip charged a used motorcycle that cost $995.65. He already owed $95.68 on the credit card. How much will the minimum monthly payment be when he is billed? $____________

(continued)

Name ______________________________ Date ____________________

24. Credit Card Buying *(continued)*

Part II

The material in the box below tells how much it will cost if you pay only part of what you owe on a credit card. Read it and answer the questions.

Hint: To find 1½% of an amount, multiply by .015.

I will pay no **finance charge** on the new balance shown on my monthly statement if such new balance is paid in full before my next billing date. Also, no **finance charge** will be assessed on any statement in which the previous charge is less than $1.00. Otherwise, I agree to pay **finance charges** at a periodic rate of 1½% per month on amounts up to $1,000 (**annual percentage rate** 18%) and a periodic rate of 1% per month on amounts in excess of $1,000 (**annual percentage rate** 12%) on the previous month's ending balance of my account.

6. Saul owed $340.00 after making a minimum payment for a month. How much will the finance charge be for the next month?

   $____________

7. Bertha still owed $525.33 after making a minimum payment for a month. How much will the finance charge be for the next month?

   $____________

8. Joe still owed $729.76 after making a minimum payment for a month. How much will the finance charge be for the next month?

   $____________

9. Terri still owed $1,252.32 after making a minimum payment for a month. How much will the finance charge be for the next month?

   $____________

Name ______________________________ Date ______________

25. Home Mortgage

Mr. Chavez sold his house and bought a more expensive one. He had to take out a $38,000.00 mortgage to cover the extra cost of the new house. After he makes his payment each month, he receives a statement like the one in the box. Read it and answer the questions.

Terms to Understand: entry date (date when the bank recorded the payment), interest (what the bank charges for lending the money), previous balance (amount owed before the most recent payment was made), new principal balance (amount owed after the latest payment was made).

YOUR RECENT PAYMENT HAS BEEN APPLIED AS SHOWN BELOW			
ENTRY DATE Mo Day	AMOUNT	TO PRINCIPAL	TO INTEREST
PREVIOUS BALANCE		37 752 87	
08 05	314 53	23 52	291 01

NEW PRINCIPAL BALANCE	TOTAL INTEREST YEAR-TO-DATE
37 729 35	2 333 05

Charles H. Chavez
2233 Old Brook Street
Wayside, Texas 77099

YOUR NEXT PAYMENT DUE	IF PAYMENT IS NOT RECEIVED BY	A LATE FEE WILL BE CHARGED AS SHOWN BELOW
09 10	09 25	31 45

1. How much was Mr. Chavez's last payment? $____________

2. How much did he owe before making his last payment? $____________

3. How much did he owe after making his last payment? $____________

4. How much of his last payment went toward paying off his loan? $____________

5. How much of his last payment was used to pay interest charges? $____________

6. If there is no late fee for sending in a payment a week late, why should he send in his payments on time?

   ____

7. Mr. Chavez borrowed a total of $38,000.00. He agreed to pay $314.53 monthly for 30 years. What will be the total amount of his payments over 30 years? $____________

8. Over 30 years, what will be the total amount of interest that Mr. Chavez will pay? $____________ (Hint: the difference between the total of all payments and the amount of the loan gives the total amount of interest.)

Name ______________________________ Date ________________

26. Car Loan

Miss Post traded in her old car and bought a better used car. The box on the right shows how much she paid. Answer the questions.

1. What was the price of the car she bought?

   $____________

2. What was the price including sales tax?

   $____________

3. How much did they allow her for trading in her old car?

   $____________

4. How much did she still owe on the old car?

   $____________

5. How much did she still owe after subtracting her trade-in from the cash price? $____________

6. How much did she need to borrow? $____________

7. How much will the loan cost her? $____________

8. How many years will she have the loan?

   $____________

9. How much will her payments be per month?

   $____________

10. How much will her payments be per year?

    $____________

11. What is the total cost of the car she is buying including sales tax, license, and loan costs? $____________

ITEM	Vehicle	$ 5,600.00	
	Sales tax	$ 336.00	
1. CASH PRICE		$5,936.00
2. DOWN PAYMENT	Previously Paid $ None	
	Paid Herewith $ None	
	CASH DOWN PAYMENT $ None	
	Gross Value of Trade-in $ 1,950.00	
	Less Lien $ None	
	TRADE-IN (Net Agreed Value) $ 1,950.00	
	TOTAL DOWN PAYMENT	$1,950.00
3. UNPAID BALANCE OF CASH PRICE (1 minus 2)		$3,986.00
4. INSURANCE (Total Gross Premium)		$ None
5. OFFICIAL FEES	License $ 110.00	
	Certificate of Title $ None	
	Registration $ Incl.	$ 110.00
6. UNPAID BALANCE (3, 4, & 5)		$4,096.00
7. DEFERRED PAYMENT		None
8. AMOUNT FINANCED (6 minus 7)		$4,096.00
9. FINANCE CHARGE		$1,474.40
10. ANNUAL PERCENTAGE RATE		15.98%
11. TOTAL OF PAYMENTS (6 & 9)		$5,570.40
Payable in 48 installments as follows 48 equal successive monthly installments of $116.05		
12. DEFERRED PAYMENT PRICE (2 & 11)		$7,520.40

12. Miss Post paid 15.98% interest on her loan. Find out the annual percentage rate a bank in your area would charge for a loan on a good used car.

    Name of bank: ____________________ **Annual percentage rate:** ____________

Name ______________________________ Date ______________

Section 5: Buying on Credit Test

1. Vanessa borrowed $1,000.00 from her credit union. After paying four monthly installments of $88.33, her balance was down to $876.00. What amount of her payments went toward finance charges? $__________

2. June owed $876.59 to a bank before she made a loan payment of $53.50. Out of her payment, $40.33 was taken out for finance charges and the rest went toward paying off the balance of the loan. How much did she owe after making the payment? $__________

3. Eric still owed a credit card company $359.99 after making the minimum monthly payment. The company charges a fee of 1½% per month on amounts that remain unpaid. How much did they charge Eric for the next month? $__________

4. Monica's credit card company offers a minimum monthly payment of $10.00 for charges totaling less than $300.00 and 4% of the total for amounts greater than $300.00. How much would Monica's minimum monthly payment be on a bill of $775.35? $__________

5. A finance charge of 1½% is applied to all overdue credit card payments under $1,000 and 1% for amounts in excess of $1,000. What is the total finance charge on an overdue amount of $1,660? $__________

6. In one year, Mrs. Albright paid $17,292 towards her $90,300 mortgage. At the end of the year, her balance was $85,553. What amount did she pay toward finance charges? $__________

7. Mr. Flint took out a mortgage in order to buy a house. He agreed to pay $876.66 per month for 30 years in order to pay off the mortgage, including interest. What is the total amount he will pay during the 30 years? $__________

8. Suzanne bought a new car for $14,113. She traded in her old car as a down payment. The dealer allowed her $4,215 for her old car. How much did Suzanne still owe after making her down payment? $__________

9. Tim wanted to use his old car, valued at $7,600, as a down payment on a new vehicle. If he has a $2,394 lien left on the car, what will be the amount applied toward his down payment? $__________

10. Olivia purchased a $12,000 car after taking out a $7,500 loan and putting $4,500 as a down payment. If she pays off her loan in three years by paying $299 monthly, what is the total cost to Olivia for the car including the loan and down payment? $__________

Section 6: Job Benefits and Income

TEACHER PAGES

27. Amount of Pay

Computational Skills
Multiplication

Mathematical Content
Fractions, Percents/Proportions/Decimals, Currency

Procedure

1. Review with less advanced students questions 1a, 1b, and 1c before assigning Part I.
2. Review with less advanced students question 4 before assigning Part II.
3. Review with less advanced students question 7a before assigning Part III.

Part I

1. a. \$47.60 (8 × \$5.95 = \$47.60)
   b. \$238.00 (40 × \$5.95 = \$238)
   c. \$1,047.20 (22 × \$47.60 = \$1,047.20)
   d. \$12,376.00 (52 × \$238 = \$12,376)
2. a. \$129.92 (8 × \$16.24 = \$129.92)
   b. \$649.60 (40 × \$16.24 = \$649.60)
   c. \$2,598.40 (20 × \$129.92 = \$2,598.40)

Part II

3. \$78.00 (\$6.50 ÷ 2 = \$3.25; \$6.50 + 3.25 = \$9.75; 8 × \$9.75 = \$78)
4. \$151.44 (\$25.24 ÷ 2 = \$12.62; \$25.24 + 12.62 = \$37.86; 4 × \$37.86 = \$151.44)
5. \$66.24 (\$14.72 ÷ 2 = \$7.36; \$14.72 + 7.36 = \$22.08; 3 × \$22.08 = \$66.24)

Part III

6. \$16.00 (40 × \$.40 = \$16)
7. \$1,560 (40 × \$.75 = \$30; \$30 × 52 = \$1,560)

28. Pay Raises and the Cost of Living

Computational Skills
Addition, Multiplication, Division

Mathematical Content
Percents/Proportions/Decimals, Currency

Procedure

1. Review with less advanced students questions 1a, 1b, and 1c, before assigning Part I.
2. Review with less advanced students questions 4, 5, and 6 before assigning Part II.
3. You may wish to point out that a person who receives a raise equal to the increase in the cost of living will be worse off than before the increase of the raise that puts him or her into a higher income-tax bracket.

Part I

1. a. \$22.80 (\$.57 × 40 = \$22.80)
   b. \$91.20 (4 × \$22.80 = \$91.20)
   c. \$1,185.60 (52 × \$22.80 = \$1,185.60)
2. a. \$.97 (\$38.80 ÷ 40 = \$.97)
   b. \$155.20 (4 × \$38.80 = \$155.20)
3. a. \$17.50 (\$70 ÷ 4 = \$17.50)
   b. \$840.00 (12 × \$70 = \$840)

Part II

4. \$599.56 (.04 × \$576.50 = \$23.06; \$576.50 + 23.06 = \$599.56)
5. \$1,319.01 (.055 × \$1,250.25 = \$68.76; \$1,250.25 + 68.76 = \$1,319.01)
6. \$472.50 (.05 × \$450.00 = \$22.50; \$450 + 22.50 =\$472.50)
7. \$1,660.68 (.039 × \$1,598.34 = \$62.34; \$1,598.34 + 62.34 = \$1,660.68)

29. Statement of Earnings

Computational Skills
Addition, Division

Mathematical Content
Tables/Graphs, Whole Numbers, Currency

Procedure

1. Have students draw up a list of specialized terms and abbreviations used on the statement of earnings and write a definition of each. Some that may appear on their list are:

pay period—period of time which the pay was earned
emp no.—employee number
rate—amount earned per unit; in this case, a unit is an hour
amount—amount of pay
deductions—amounts an employee has withheld from his or her paycheck for specific purposes
health ins.—health insurance
401(k) plan—retirement plan that provides income after an employee retires; dollars invested through an outside company in funds chosen by the employer
ytd—year-to-date balance, or total for the year, including this paycheck
gross—amount of earnings before deductions
fit w/h—federal income tax withholding; money kept back for federal income tax
fica—social security taxes; note; initials stand for "federal income contributions act."
sit w/h—state income tax withholding
net pay—amount of pay after deductions; take-home pay
federal tax—amount employee pays to federal government
state tax—amount employee pays to state government

2. Point out to students that the bottom line of figures are for the year to date; the next-to-the-bottom line provides information on the current pay period.
3. Students of varying abilities may need help with the entire exercise.
    1. 80 hours
    2. $1,000.00
    3. $100 (80 ÷ 8 = 10; $1,000 ÷ 10 = $100)
    4. $739.05
    5. $73.91 ($739.05 ÷ 10 = $73.91)
    6. $1.40 ($13.97 ÷ 10 = $1.40)
    7. $168.76 (Note: ytd = year-to-date balance)
    8. $774.44
    9. $65.30
    10. $98.86 ($90.13 + $8.73 = $98.86) Note: fit w/h = federal income tax withholding; sit w/h = state income tax withholding.

30. Personal/Vacation Leave

Computational Skills
Addition, Multiplication

Mathematical Content
Tables/Graphs, Whole Numbers, Fractions, Percents/Proportions/Decimals

Procedure

1. Review with less advanced students items 5, 6, and 10 before assigning Part II.
2. Students may solve the problems in the second part by (1) multiplying the months worked by the fractions given or (2) converting the fractions to decimals and then multiplying.
    1. None (He must be employed for a full year before being given vacation credit.)
    2. 13
    3. 6
    4. 9
    5. 2
    6. a. 3 ($4 \times \frac{3}{4} = 1 \times 3 = 3$ OR $4 \times .75 = 3$)
       b. 1 ($4 \times \frac{1}{4} = 1$)
    7. 18 ($24 \times \frac{3}{4} = 6 \times 3 = 18$ OR $24 \times .75 = 18$)
    8. 7 ($48 \times \frac{1}{4} = 12$; $12 - 5 = 7$)
    9. 9 ($12 \times \frac{3}{4} = 3 \times 3 = 9$ OR $12 \times .75 = 9$)
    10. 9 (Same as number 8.)

11. 15 ($12 \times 1\frac{1}{4} = 12 \times 5/4 = 3 \times 5 = 15$ OR $12 \times 1.25 = 15$)
12. a. 3 ($12 \times \frac{1}{4} = 3$)
    b. 3 (Same as in a.)

31. Sick Leave

Computational Skills
Multiplication, Division

Mathematical Content
Percents/Proportions/Decimals

Procedure

1. Review the multiplication of whole numbers by decimals using item 1.
2. Review with less advanced students questions 1, 2, 4, and 7 before assigning the exercise.
3. You may also want to discuss the benefits of long- and short-term disability: company-paid percentage of salary for employees off for extended periods of time due to illnesses, family leave, etc.
    1. 0 (She hasn't worked seven weeks yet.)
    2. 7.43 ($52 \div 7 = 7.43$)
    3. 14.86 ($52 \times 2 = 104$; $104 \div 7 = 14.86$)
    4. 5.25 ($.583 \times 9 = 5.247 = 5.25$)
    5. 13.99 ($.583 \times 24 = 13.992 = 13.99$)
    6. 20.99 ($.583 \times 36 = 20.988 = 20.99$)
    7. No (Credit is not given until after six months on the job.)
    8. 19.99 ($.833 \times 24 = 19.992 = 19.99$; this company probably would credit with 10 days for a full year, giving 20 days for two full years.)
    9. 29.99 ($.833 \times 36 = 29.988 = 29.99$)

32. Purchase Order

Computational Skills
Addition, Multiplication

Mathematical Content
Percents/Proportions/Decimals, Currency

Procedure

1. You may want to bring a mail-order catalog for students to look through. Some items may cost less when ordered in bulk.
2. Point out that in addition to the cost of materials ordered, some companies may charge sales tax, postage and handling costs, or other processing charges.

The form should be filled out as shown here.

				Date Needed	ASAP	
	WANTED		DESCRIPTION Give accurate description and purpose of intended use. Indicate catalog reference whenever possible	ESTIMATED COST		
	Quantity	Unit or Measure		Per Unit	Total	
1	4	doz	Pencils, standard #2	1.25	5	00
2	2	doz	Pens, ballpoint—fine point, black	3.44	6	88
3	3	doz	Pens, ballpoint—medium point, black	3.25	9	75
4	6	ea	Printer inkjet cartridges	28.57	171	42
5	4	pair	Scissors	5.60	22	40
6	8	blt	Soap, ditto liquid	2.00	16	00
7	12	box	Staples, standard	.95	11	40
8	7	ea	Stapling machine, desk model	9.99	69	93
9	24	roll	Facsimile paper	1.77	42	48
10	6	ea	Binder, two-piece pressboard, black	3.33	19	98
11	6	pkg	Binder, divisional sheets, 81/2 x 11"	1.60	9	60
12	12	pkg	Binder, filler sheets, 81/2 x 11"	1.55	18	60
13	2	ea	File cabinet—four drawer	119.95	239	90
14						
15						
				SUB-TOTAL	643	34
				SALES TAX	32	17
				OTHER CHARGES	0	
				TOTAL	675	51

DEPARTMENT COMPLETE THE FOLLOWING:

I hereby certify that the goods and/or services requested hereon are necessary to properly implement authorized programs under my supervision.

Janet Deveny

Department Head or other authorized representative

Ordered by Tom Jones Ext. 3711

Deliver to Main Building, Room A

Do Not Write in This Block
BUSINESS OFFICE DATA

________ No. ________

Completed by ________

Date ________

33. Health Benefits

Computational Skills
Subtraction, multiplication

Mathematical Content
Tables/Graphs, Percents/Proportions/Decimals, Currency

Procedure

Discuss the differences between major medical and Health Maintenance Organizations (HMO's). HMO's limit the service provided and the doctors who provide service. Major medical has fewer restrictions, etc.

Discuss reasons for differences in rates between major medical plans and HMO's. Factors involved may include deductibles, co-pays, hospital stay coverage, dental coverage, out-of-area benefits, immunizations covered, types of routine exams covered, mental health and substance abuse benefits, physical therapy coverage, lab fees, X-ray fees, etc.

1. a. $76.65
   b. $919.80
   c. $85.35
   d. $1,024.20
2. a. $208.70
   b. $56.55 ($265.25 – 208.70 = $56.55)
   c. $53.25 ($265.25 – 212.00 = $53.25)
   d. $639.00 ($12 × $53.25 = $639.00)
3. a. $11.10
   b. $0

34. Life Insurance

Computational Skills
Multiplication, Division

Mathematical Content
Tables/Graphs, Percents/Proportions/Decimals, Currency

Procedure

1. Define the term *life insurance*—a benefit providing payment of a set amount to the family in the event of the employee's death.

2. Make sure students know to multiply the annual salary by 1, 2, or 3, depending on which option employees choose. Students must also divide the salary by 1,000 since rates are given per $1,000.00.

   1. a. $0.05
      b. $1.40 (28 × $.05 = $1.40)
   2. a. $37.17 (63 × .59 = $37.17)
      b. $446.04 ($37.17 × 12 = $446.04)
   3. $4.40 (22 × 2 = 44 × .1 = $4.40)
   4. a. $13.50 (45 × 3 = 135 × .1 = $13.50)
      b. $162.00 ($13.50 × 12 = $162.00)
      c. $356.40 (135 × .22 = $29.7 × 12 = $356.40)
   5. a. $68.12 (52 × $1.31 = $68.12)
      b. $61.36 (52 × 2 = 104 × .59 = $61.36)

35. Travel Expense Form

Computational Skills
Addition, Subtraction

Mathematical Content
Tables/Graphs, Currency

Procedure

1. Explain that not all companies give a set dollar amount to an employee before a trip. Some may not give any; they may only reimburse after getting receipts. They may also pay for everything up front, with a company credit card.
2. You may wish to have students check their own work by seeing if the sum of the column totals equals the sum of the line (row) totals.
3. Less advanced students may have difficulty with the five entries just below the space for total expenses. You may wish to have them independently compute the total expenses and help them with the last five entries.

Travel Expense Form

Dates	SUNDAY 9/10	MONDAY 9/11	TUESDAY 9/12	WEDNESDAY 9/13	THURSDAY /	FRIDAY /	SATURDAY /	Total each line
Breakfast (include tip)	5:25	4:60	3:00	5:17				18:02
Lunch (include tip)	5:50	8:20	7:65	5:50				26:85
Dinner (include tip)	18:00	9:75	10:50					38:25
Hotel/motel room	95:40	95:40	95:40					286:20
Limousine, taxi or bus	13:75	8:75	9:00	11:75				43:25
Plane or train fares (1)								
Travel tickets furnished by IR	523:32							523:32
Automobile 25¢ per mile	11:25			11:25				22:50
Parking				21:00				21:00
Tolls								
Gratuities	5:00							5:00
Miscellaneous: (2)								
Xerox		3:50						3:50
Total each column	677:47	130:20	125:55	54:67				**Total expenses** 987:89

(2) Explain items under Miscellaneous and any unusual items that may be questioned.

Xeroxed 35-page report for customer

(1) If first-class fare, explain why.

Less advances cash	(400:00)
Travel tickets furnished by IR	(532:32)
Total advances	(932:32)
Due traveler	(55:57)
Due IR	(00:00)

Section 6 Test

1. \$104.40 (\$11.60 × 1.5 = \$17.40 × 6 = \$104.40)
2. \$22 (40 × \$0.55 = \$22)
3. \$3,090 (\$154.50 × 20 = \$3,090)
4. 9 (¾ × 12 = 9)
5. 8 (.667 × 12 = 8)
6. \$78.84 (\$23.95 × 3 = \$71.85 + \$6.99 = \$78.84)
7. \$133.32 (\$99.88 – \$88.77 = \$11.11 × 12 = \$133.32)
8. \$15.91 (\$37,000 ÷ 1,000 = 37 × \$.043 = \$15.91)
9. \$14 (56 × \$0.25 = \$14)
10. \$192.75 (\$442.75 – \$250.00 = \$192.75)

Name ______________________________ Date ______________

27. Amount of Pay

Part I

Answer the following questions.

1. a. Jonas earns $5.95 per hour. How much does he earn in an eight-hour day? $__________

   b. How much does Jonas earn in a 40-hour week? $__________

   c. How much does Jonas earn in a month with 22 workdays? $__________

   d. How much does Jonas earn in a year (52 weeks)? $__________

2. a. Carola earns $16.24 per hour. How much does she earn in an eight-hour workday? $__________

   b. How much does Carola earn in a 40-hour week? $__________

   c. How much does Carola earn in a month with 20 workdays? $__________

Part II

People who work more than eight hours in a day or 40 hours in a week often earn overtime pay. Overtime pay in all the following questions is 1½ times regular pay.

3. Daniel's regular pay is $6.50 per hour. One week he worked eight hours overtime. How much did he earn in overtime for that week? $__________

4. Sherri's regular pay is $25.24 per hour. One week she worked four hours overtime. How much did she earn for overtime that week? $__________

5. Marci's regular pay is $14.72 per hour. One day she worked three hours overtime. How much did she earn for overtime that day? $__________

Part III

People who work nights instead of days often receive extra pay called a "differential."

6. Saul's regular pay is $8.67 per hour. The night shift differential where he works is 40 cents per hour. How much more will Saul earn per 40-hour week by working nights instead of days? $__________

7. Lila's regular pay is $16.60 per hour. The night shift differential where she works is 75 cents per hour. How much more will Lila earn per year (52 weeks) by working nights instead of days? $__________

Name ______________________________ Date ______________

28. Pay Raises and the Cost of Living

Part I

Answer the following questions about pay raises.

1. a. All bus drivers in a city received a pay raise of $.57 per hour. How much more did each bus driver receive per 40-hour week? $____________

   b. In a month with exactly four weeks, how much more did each bus driver receive? $____________

   c. How much more did each bus driver receive per year (52 weeks)? $____________

2. a. Each machine technician in a plant received a pay raise of $38.80 per 40-hour work week. How much more did each one receive per hour? $____________

   b. In a month with exactly four weeks, how much more did each machine operator receive? $____________

3. a. Each state employee received a raise of $70.00 per month. In a month with exactly four weeks, how much more did each state employee receive per week? $____________

   b. How much more did each state employee receive per year? $____________

Part II

Most years, the cost of things we buy (cost of living) goes up. Answer the following questions about the cost of living and pay raises. Round off your answers to two decimal places. If the numeral in the third place in your answer is 5 or larger, round up.

Hint: To find 4% of an amount, multiply by .04. To find 5.5% of an amount, multiply by .055.

4. Anna's salary was $576.50 per week. She received a 4% pay raise. What was her salary after the raise?

   $____________

5. Sam's salary was $1,250.25 per week. He received a 5.5% pay raise. What was his salary after the raise?

   $____________

6. In a recent year, the cost of living went up 5%. A person earning $450.00 per week near the beginning of the year will need how much per week near the end of the year in order to keep up with the cost of living?

   $____________

7. In a recent year, the cost of living went up 3.9%. A person earning $1,598.34 per month near the beginning of the year will need to earn how much per month near the end of the year in order to keep up with the cost of living? $____________

Name ______________________ Date ______________

29. Statement of Earnings

In the box is a statement of earnings that John Baccia received with his paycheck. A "unit" on this statement is one hour. Be sure you know the meanings of all the other terms and abbreviations on the statement before answering the questions.

Hint: Compute your answers to three decimal places and round off to two. If the third decimal place is 5 or more, round up.

ABC International Company	STATEMENT OF EARNINGS		
emp. no 608	employee name John D. Baccia		check no. 322777
from 07-31	pay period	to 8-14	social security no. 009-34-2390

earnings				deductions		
description	units	rate	amount	description	amount	ytd
REGULAR	80.00	12.50	1,000.00	Health Ins.	13.97	168.76
				Social Security	65.30	562.40
				Federal Tax	90.13	1,080.28
				State Tax	8.73	99.81
				401k Plan	82.82	774.44
total earnings		$	1,000.00	deductions	260.95	2,685.69

current and year to date

gross	fit w/h	fica		sit w/h	city	other	net pay
1,000.00	90.13	65.30		8.73		96.79	$ 739.05
13,000.00	1,080.28	562.40		99.81		943.20	$10,314.31

1. How many hours did John work during the pay period? ____________

2. How much did John earn during the pay period? $____________

3. John worked eight hours per day. How much did he earn per day during the pay period? $____________

4. How much money did he take home for the pay period? $____________

5. How much were his earnings per day during the pay period after all deductions and taxes were withheld? $____________

6. How much did health insurance cost him per day? $____________

7. How much was withheld from January 1 through August 14 for health insurance? $____________

8. How much has been invested in his retirement plan from January 1 through August 14? $____________

9. How much was withheld during the pay period for Social Security? $____________

10. How much was withheld during the pay period for income taxes? $____________

Name ______________________________ Date ______________

30. Personal/Vacation Leave

The personal and vacation leave policies of two companies are described in the box below and on page 73. Answer the questions under each box.

ABC Company

Personal Leave Policy:
Each employee earns two days of paid personal leave for each year worked. During the first year of employment, employees must work at least seven full months to receive paid leave.

Vacation Leave Policy:
Each employee earns one day of paid vacation leave per month worked. Vacation days earned during the first year of employment are not available for use until the first day of the second year of employment.

1. After being employed for 10 months, Greg wants to take a vacation. How many paid vacation days can he take?

   ____

2. After being employed for 13 months, Jane wants to take a vacation. How many paid vacation days can she take?

   ____

3. Randy returned to work from vacation on October 1 of this year. He used all of his vacation days on the vacation. How many paid vacation days will he have on April 1 of next year?

   ____

4. Diana returned to work from vacation on November 1 of this year. She used all of her vacation days on the vacation. How many paid vacation days will she have on August 1 of next year?

   ____

5. Luwanda was hired four months into the new year. Employees earn ¼ of a day personal time for each month worked. How many personal days may she take?

   ____

(continued)

Name ______________________________ Date ______________

30. Personal/Vacation Leave *(continued)*

> **XYZ Company**
>
> **Personal Leave Policy:**
> Each employee is credited with $\frac{1}{4}$ of a day personal time for each month worked. No employee may accumulate more than 10 days paid personal leave credits.
>
> **Vacation Leave Policy:**
> Through the fifth year, each employee is credited with $\frac{3}{4}$ of a day paid vacation leave for each month worked.
> Beginning on the first day of the sixth year of employment, each employee is credited with $1\frac{1}{4}$ days paid vacation leave for each month worked. No employee may accumulate more than 25 days of paid vacation leave credits.

6. a. After being employed for four months, how many paid vacation days has a new employee earned?

   ____

   b. How much personal time has the employee earned? ______________

7. After being employed for two years, Michael wants to take a vacation. How many paid vacation days can he take?

   ____

8. After being employed for four years, Sarah has taken only five personal days. How many paid personal days can she take?

   ____

9. How many paid vacation days does an employee earn during the first year of employment?

   ____

10. How many paid vacation days does an employee earn during the third year of employment?

    ____

11. How many paid vacation days does an employee earn during the eleventh year of employment?

    ____

12. a. How many paid personal days does an employee earn during the second year of employment?

    ____

    b. How many personal days does an employee earn during the fifteenth year?

    ____

Name ______________________ Date ______________

31. Sick Leave

The sick leave policies of three companies are described in the boxes. Answer the questions under each box. Round off your answers to two decimal places.

> **Solo Company Sick Leave Policy**
> Each employee shall earn one day of paid sick leave for each seven weeks worked.

1. Penny has been on the job for five weeks. How many days of sick leave has she earned? ____________
2. Skip has been on the job for one year. How many days of sick leave has he earned? ____________
3. Tom has been on the job for two years. How many days of sick leave has he earned? ____________

> **Downhill Company Sick Leave Policy**
> Each employee shall earn .583 days of paid sick leave for each full month worked.

4. Martha has been on the job for nine months. How many days of sick leave has she earned?

   ____

5. Sarah has been on the job for two years. How many days of sick leave has she earned? ____________
6. Terry has been on the job for three years. How many days of sick leave has she earned? ____________

> **Uptown Company Sick Leave Policy**
> Each employee shall earn .833 days of paid sick leave for each full month worked. Sick leave shall not be credited to new employees until after the first full six months of employment.

7. Ralph took two days off because he was sick after being on the job for three months. Was he paid for the two days? ____________
8. Rafe has been on the job for two years. How many days of sick leave has he earned? ____________
9. Sadie has been on the job for three years. How many days of sick leave has she earned? ____________

Name ______________________ Date ______________

32. Purchase Order

This is an order for materials placed by Tom Jones, who supervises the office staff of a medium-sized company. The cost per unit for each item is given. The sales tax rate is five percent. Complete the order form by filling in the last column.

Date Needed ASAP

	WANTED		DESCRIPTION Give accurate description and purpose of intended use. Indicate catalog reference whenever possible	ESTIMATED COST	
	Quantity	Unit or Measure		Per Unit	Total
1	4	doz	Pencils, standard #2	1.25	
2	2	doz	Pens, ballpoint—fine point, black	3.44	
3	3	doz	Pens, ballpoint—medium point, black	3.25	
4	6	ea	Printer inkjet cartridges	28.57	
5	4	pair	Scissors	5.60	
6	8	blt	Soap, ditto liquid	2.00	
7	12	box	Staples, standard	.95	
8	7	ea	Stapling machine, desk model	9.99	
9	24	roll	Facsimile paper	1.77	
10	6	ea	Binder, two-piece pressboard, black	3.33	
11	6	pkg	Binder, divisional sheets, 81/2 x 11"	1.60	
12	12	pkg	Binder, filler sheets, 81/2 x 11"	1.55	
13	2	ea	File cabinet—four drawer	119.95	
14					
15					
				SUB-TOTAL	
				SALES TAX	
				OTHER CHARGES	0
				TOTAL	

DEPARTMENT COMPLETE THE FOLLOWING:

I hereby certify that the goods and/or services requested hereon are necessary to properly implement authorized programs under my supervision.

Janet Deveny
Department Head or other authorized representative

Ordered by Tom Jones Ext. 3711

Deliver to Main Building, Room A

Do Not Write in This Block
BUSINESS OFFICE DATA

__________ No. __________

Completed by __________

Date __________

Name ______________________________ Date ______________

33. Health Benefits

In the box are the monthly rates for health plans that employees at one company may choose from. The employee must pay the amount offered in the table. Whatever health costs are remaining are paid by the employer. Answer the following questions.

	FAMILY HEALTH		FIRST FARWEST		FOUNDATION HEALTH HMO		FRENCH HEALTH HMO	
	PLAN CODE	MONTHLY RATE	PLAN CODE	MONTHLY RATE	PLAN CODE	MONTHLY RATE	PLAN CODE	MONTHLY RATE
EMP. ONLY	591	$87.50	621	$62.00	841	$39.50	2171	$28.48
EMP. & 1 DEP.	592	208.70	622	162.00	842	76.65	2172	78.95
EMP. & 2 OR MORE DEP.	593	265.25	623	212.00	843	132.80	2173	121.70

1. a. Jennifer chose Foundation Health HMO. She has one dependent. How much does she personally pay per month?

   $______________

   b. How much does Jennifer pay per year? $______________

   c. How much more would Jennifer spend per month if she switched to First Farwest? $______________

   d. How much more would Jennifer spend per year if she switched to First Farwest? $______________

2. a. Fernando chose Family Health. He has one dependent. How much does he personally pay per month?

   $______________

   b. Fernando and his wife will have a child later this year. How much more per month will it cost him personally when he has two dependents instead of just one?

   $______________

   c. After he has two dependents, how much would Fernando save per month by switching to First Farwest?

   $______________

   d. With two dependents, how much would Fernando save per year by switching to First Farwest?

   $______________

3. a. Gilda currently is paying to insure herself and two children. She wants to switch from Foundation Health to French Health HMO. How much will this save her per month?

   $______________

   b. Gilda is expecting another child. How much more per month will it cost her when she has four dependents instead of just three?

   $______________

Name ________________________________ Date ________________

34. Life Insurance

Upscale Enterprises Inc. covers each of its employees under a $100,000 life insurance policy. The company also offers employees the option of purchasing additional life insurance at a discounted rate. Employees may purchase extra life insurance equal to one, two, or three times their current annual salary based on the rates in the following table.

Age (years)	Rate/$1,000 (per mo.)
< 30	0.05
31 – 40	0.10
41 – 50	0.22
51 – 60	0.59
> 61	1.31

1. a. Renee, who is 29 years old, wants to buy additional life insurance equal to one times her annual salary of $28,000. How much per $1,000 will she be charged each month?

   $______________

   b. What will her monthly cost be? $______________

2. a. Arnold, who earns $63,000 each year, wants to buy life insurance equal to one times his salary. He is 57.

   What will he pay each month? $______________

   b. What will his yearly cost be for this benefit? $______________

3. Trish, likewise, requested a monthly deduction for life insurance. She is 34 and earns $22,000 a year. If the amount being deducted covers twice her salary, how much is Trish having deducted?

   $______________

4. a. Francine, 40, wants extra coverage equal to three times her salary of $45,000.

   What will her monthly cost be? $______________

   b. What is her yearly cost? $______________

   c. Next year, what will her yearly cost be? $______________

5. a. When Mr. Boyer turned 61, he decided to decrease the amount being taken from his paycheck for life insurance from two times his salary of $52,000 to just one times his salary. What is his monthly charge now?

   $______________

   b. What was his monthly charge last year? $______________

Name ______________________ Date ______________

35. Travel Expenses Form

Judy Low works for International Radio (IR). She was asked to travel to another city on business. IR gave her an airline ticket that cost $523.32 and $100.00 per day in advance of the trip to help pay her expenses. She drove 90 miles round trip to and from the airport. Finish filling out the travel expense for her.

Hints: Enter the amount of cash she was given to the right of the words "Less advances: cash." Enter the cost of the airline tickets she was given to the right of the words. "Travel tickets furnished by IR." If the Total expenses are greater than the Total advances, IR owes Judy money for the trip. That is, there is an amount "Due traveler."

Travel Expense Form

Dates	SUNDAY 9/10	MONDAY 9/11	TUESDAY 9/12	WEDNESDAY 9/13	THURSDAY /	FRIDAY /	SATURDAY /	Total each line
Breakfast (include tip)	5.25	4.60	3.00	5.17				
Lunch (include tip)	5.50	8.20	7.65	5.50				
Dinner (include tip)	18.00	9.75	10.50					
Hotel/motel room	95.40	95.40	95.40					
Limousine, taxi or bus	13.75	8.75	9.00	11.75				
Plane or train fares (1)								
Travel tickets furnished by IR	523.32							
Automobile 25¢ per mile	11.25			11.25				
Parking				21.00				
Tolls								
Gratuities	5.00							
Miscellaneous: (2)								
Xerox		3.50						
Total each column								Total expenses

(2) Explain items under Miscellaneous and any unusual items that may be questioned.

Xeroxed 35-page report for customer

(1) If first-class fare, explain why.

Less advances cash	( )
Travel tickets furnished by IR	( )
Total advances	( )
Due traveler	( )
Due IR	( )

Name ______________________________ Date ______________

Section 6: Job Benefits and Income Test

1. Ron's regular pay is $11.60 per hour. One week, he worked six hours overtime. His overtime pay is 1½ times his regular pay. How much did he earn for overtime that week? $__________

2. All mechanics in a factory received a $0.55 pay raise per hour. How much more did each mechanic receive per 40-hour week? $__________

3. James earns $154.50 per day. His paycheck for 20 days of work should show how much of his total earnings? $__________

4. Each employee at a company earns three quarters of a day paid vacation for each month worked. After working for a year, how many paid vacation days does each employee have? __________

5. Each employee at a company earns .667 days of paid sick leave for each month worked. How many paid sick days does each employee earn in one year? __________

6. Sal put in a purchase order for three inkjet cartridges for his laser printer. The cost for each cartridge is $23.95. What is the total cost of Sal's order if he must also pay a $6.99 shipping charge on his order? $__________

7. If the ABC insurance company charges $99.88 per month for health insurance and the XYZ company charges $88.77 per month, how much more will you save per year by buying insurance from XYZ instead of ABC? $__________

8. Last year, Darcy opted to pay an additional $0.43 per $1,000 of her $37,000 yearly salary to increase the value of her life insurance policy. How much of her salary did Darcy put toward life insurance? $__________

9. For business traveling, Gina's company pays $0.25 per mile traveled when driving by car. How much would the company owe Gina if she traveled 56 miles on a business trip? $__________

10. Lance went on a business trip for his company. Before he left, his company advanced him $250.00 in cash for the trip. The trip cost him $442.75. How much does his company owe him? $__________

Section 7: Taxes

TEACHER PAGES

36. Sales Tax Computations

Computational Skills
Subtraction, Multiplication

Mathematical Content
Percents/Proportions/Decimals, Currency

Procedure

Review with less advanced students questions 1a, 1b, and 1c before assigning the exercise.

1. a. \$270.00 (.06 × \$4,500.00 = \$270.00)
   b. \$180.00 (.04 × \$4,500.00 = \$180.00)
   c. \$90.00 (\$270.00 – \$180.00 = \$90.00)
2. a. \$7.80 (.06 × \$130.00 = \$7.80)
   b. \$6.50 (.05 × \$130.00 = \$6.50)
   c. \$1.30 (\$7.80 – \$6.50 = \$1.30)
3. a. \$22.75 (.065 × \$350.00 = \$22.75)
   b. \$19.25 (.055 × \$350.00 = \$19.75)
4. a. \$2.40 (.06 × \$40.00 = \$2.40)
   b. \$0.80 (.04 × \$40.00 = \$1.60; \$2.40 – \$1.60 = \$.80)
5. a. \$172.00 (.04 × \$4,300.00 = \$172.00)
   b. \$43.00 (.05 × \$4,300.00 = \$215.00; \$215.00 – \$172.00 = \$43.00)

37. Federal Income Tax

Computational Skills
Addition, Subtraction

Mathematical Content
Tables/Graphs, Currency

Procedure

1. Have students read the entire 1040EZ form before starting this exercise. Go over with them each individual line on the form. Make sure they understand all unfamiliar terms or concepts (IRS, W-2 form, taxable interest income, unemployment compensation, adjusted gross income, claim as a dependent, taxable income, earned income credit, direct deposit, etc.)
2. Then have students read the introductory paragraph to the exercise and complete as much of the tax return form as possible. Students who are less advanced or unfamiliar with tax returns may need help.
3. Let students know that there are different types of forms and schedules that have to be completed when filing a tax return, depending on a person's (or business's) income and status situation. The 1040EZ is the most basic of these forms.

4. Point out that information for filing tax returns for the past year is supplied to them by mail in January. W-2 forms come from their employer, interest income information is sent from the banks or companies where they invest their money, and tax return forms are sent from the government (or can be picked up at nearby public locations).
5. Remind students that when sending in tax return information, a copy of their W-2 must be attached to the return before mailing it back to the IRS.
6. When completing this exercise, it is important to read all directions carefully and thoroughly. Encourage students to check their addition and subtraction and to look over their work again before they finish. This is a very important practice.

Form **1040EZ** Department of the Treasury—Internal Revenue Service
Income Tax Return for Single and Joint Filers With No Dependents (U) **1997** OMB No. 1545-0675

Use the IRS label here

Your first name and initial: Roger A. Last name: Hausman
If a joint return, spouse's first name and initial — Last name
Home address (number and street). If you have a P.O. box, see page 7. — Apt. no.: 386 Rocky Ridge
City, town or post office, state, and ZIP code. If you have a foreign address, see page 7.: Oakville, MN 55483

Your social security number 012 34 5678
Spouse's social security number

Presidential Election Campaign (See page 7.)
Note: *Checking "Yes" will not change your tax or reduce your refund.*
Do you want $3 to go to this fund? ▶ Yes No
If a joint return, does your spouse want $3 to go to this fund? ▶ Yes No

Dollars Cents

Income
Attach Copy B of Form(s) W-2 here. Enclose but do not attach any payment with your return.

1 Total wages, salaries, and tips. This should be shown in box 1 of your W-2 form(s). Attach your W-2 form(s). 1
2 Taxable interest income. If the total is over $400, you cannot use Form 1040EZ. 2
3 Unemployment compensation (see page 9). 3
4 Add lines 1, 2, and 3. This is your **adjusted gross income.** If under $9,770, see page 9 to find out if you can claim the earned income credit on line 8a. 4

Note: *You must check Yes or No.*
5 Can your parents (or someone else) claim you on their return?
Yes. Enter amount from worksheet on back. **No.** If **single,** enter 6,800.00. If **married,** enter 12,200.00. See back for explanation. 5
6 Subtract line 5 from line 4. If line 5 is larger than line 4, enter 0. This is your **taxable income.** ▶ 6

Payments and tax
7 Enter your Federal income tax withheld from box 2 of your W-2 form(s). 7
8a **Earned income credit** (see page 9).
b Nontaxable earned income: enter type and amount below. Type $ 8a
9 Add lines 7 and 8a. These are your **total payments.** 9
10 **Tax.** Use the amount on **line 6** to find your tax in the tax table on pages 20–24 of the booklet. Then, enter the tax from the table on this line. 10

Refund
Have it directly deposited! See page 13 and fill in 11b, 11c, and 11d.
11a If line 9 is larger than line 10, subtract line 10 from line 9. This is your **refund.** 11a
▶ b Routing number
▶ c Type: Checking Savings d Account number

Amount you owe
12 If line 10 is larger than line 9, subtract line 9 from line 10. This is the **amount you owe.** See page 13 for details on how to pay. 12

I have read this return. Under penalties of perjury, I declare that to the best of my knowledge and belief, the return is true, correct, and accurately lists all amounts and sources of income I received during the tax year.

Sign here ▶ Keep copy for your records.
Your signature — Spouse's signature if joint return
Date — Your occupation — Date — Spouse's occupation

For Official Use Only

For Privacy Act and Paperwork Reduction Act Notice, see page 18. Cat. No. 14090A 1997 Form 1040EZ

38. State and Local Taxes

Computational Skills
Addition, Subtraction, Multiplication

Mathematical Content
Percents/Proportions/Decimals, Currency

Procedure

Point out to students that in these examples taxable income for a person is less than the total wages because taxable income is determined after making allowances for exemptions and deductions.

1. a. \$198.76 (\$19,876 × .01 = \$198.76)
   b. \$506.31 (\$16,877 × .03 = \$506.31)
   c. \$705.07 (\$198.76 + 506.31 = \$705.07)
2. a. \$849.14 (\$42,457 × .02 = \$849.14)
   b. \$1263.12 (\$31,578 × .04 = \$1263.12)
3. \$220.00 (\$28,000 – \$26,000 = \$2,000; \$2,000 × .11 = \$220.00)
4. a. \$2,100 (\$14,000 × .15 = \$2,100)
   b. \$840 (\$14,000 × .06 = \$840)
   c. \$11,060 (\$14,000 – 2,100 – 840 = \$11,060)
5. a. \$1780 (\$10,000 x .15 = \$1,500; \$4,000 × .07 = \$280; \$1,500 + 280 = \$1,780)
   b. \$12,220 (\$14,000 – 1,780 = \$12,220)

39. Use Tax

Computational Skills
Addition, Multiplication

Mathematical Content
Tables/Graphs, Currency

Procedure

1. Explain what a use tax is. Some states charge a percent tax on items purchased for personal use from companies and/or services not located within the state of residence. (Mail-order companies, television-based home shopping networks, Internet orders, repair on maintenance services located out-of-state—these are subject to use taxes.) A use tax is like a sales tax on items. Use taxes are usually paid yearly as part of a person's state income tax return (if one needs to be filed).
2. Ask students to think of some other types of goods and services that may be subject to use taxes.
3. Review with students of varying abilities how to change percents to decimals and how to multiply by percents.

   1. \$43.44 (\$725 × .06 = \$43.44)
   2. \$0 (Note: use tax for goods valued > \$50 only)
   3. \$56.38 (\$939.68 × .06 = \$56.38)

4. a. \$19.20 (\$480 × .04 = \$19.20)
   b. \$9.60 (\$480 × .06 = \$28.80 – \$19.20 = \$9.60)
5. a. \$3.99 (\$79.80 × .05 = \$3.99)
   b. \$.80 (\$79.80 × .06 = \$4.79 – \$3.99 = \$.80)
6. a. 4½ %
   b. \$25.45 (\$565.65 × .045 = \$25.45)

Section 7 Test

1. \$2.34 (\$19.50 × 2 = \$39 × .06 = \$2.34)
2. \$13.15 (\$239 × .055 = \$13.15)
3. \$0.92 (\$46 × .05 = \$2.30; \$46 × .07 = \$3.22 – \$2.30 = \$0.92)
4. \$8,990.52 (\$32,109 × .28 = \$8,990.52)
5. \$1,087.80 (\$1,006.33 + \$81.47 = \$1,087.80)
6. \$1,857.97 (\$1,342.18 + \$515.79 = \$1,857.97)
7. \$7,650 (\$9,000 × .15 = \$1,350; \$9,000 – \$1,350 = \$7,650)
8. \$319.55 (\$21,303 × .015 = \$319.55)
9. \$22.95 (\$459 × .05 = \$22.95)
10. \$7.67 (\$117.95 × .065 = \$7.67)

Name ______________________________ Date ______________

36. Sales Tax Computations

Answer the following questions.

Hint: To find 6% of an amount, multiply by .06.

1. a. How much is the sales tax on a motorcycle that costs $4,500.00 in a state with a 6% sales tax?

   $__________

   b. How much is the sales tax on the motorcycle in a state with a 4% sales tax? $__________

   c. How much more sales tax would you pay if the rate was 6% instead of 4%? $__________

2. a. How much is the sales tax on a 10-speed bicycle that costs $130.00 in a state with a 6% sales tax?

   $__________

   b. How much is the sales tax on the bicycle in a state with a 5% sales tax? $__________

   c. How much more sales tax is paid on the bicycle with a 6% tax rate instead of a 5% tax rate?

   $__________

3. a. How much is the sales tax on a stereo that costs $350.00 in a state with a 6½% sales tax?

   $__________

   b. How much is the sales tax on the stereo in a state with a 5½% sales tax? $__________

4. a. How much is the sales tax on a bottle of perfume that costs $40.00 in a state with a 6% sales tax?

   $__________

   b. How much less sales tax is paid on the perfume in a state with a 4% sales tax compared with a state with a 6% sales tax? $__________

5. a. If you spend $4,300.00 in a year on taxable items in a state with a 4% sales tax, how much will you pay in sales taxes for the year? $__________

   b. The governor wants to raise the sales tax to 5%. If she wins, how much more will you pay in sales tax for the year? $__________

Name ______________________ Date ______________

37. Federal Income Tax

Roger A. Hausman is using the 1040EZ Federal Income Tax Return form to file his income tax for the previous year. He is single and claiming himself as a dependent. His total salary from last year was $18,384.23. He earned a total of $262.16 in interest on both his savings account and stocks. A total of $1,588.60 was withheld for tax purposes. He was employed the entire past year but has no additional earned income credit. According to the tax table, Roger should pay $1,774.00 in taxes based on his previous year's income. He does not want to donate any money toward the presidential election campaign. Using this information, fill in the 1040EZ form to see if Roger is due a refund or if he owes the government.

Form **1040EZ** — Department of the Treasury—Internal Revenue Service
Income Tax Return for Single and Joint Filers With No Dependents (U) 1997 OMB No. 1545-0675

Use the IRS label here

Your first name and initial: Roger A. — Last name: Hausman
If a joint return, spouse's first name and initial — Last name
Home address (number and street). If you have a P.O. box, see page 7. — Apt. no.: 386 Rocky Ridge
City, town or post office, state, and ZIP code. If you have a foreign address, see page 7.: Oakville, MN 55483

Your social security number: 012 34 5678
Spouse's social security number:

Presidential Election Campaign (See page 7.)
Note: *Checking "Yes" will not change your tax or reduce your refund.*
Do you want $3 to go to this fund? ▶ Yes ☐ No ☐
If a joint return, does your spouse want $3 to go to this fund? ▶ Yes ☐ No ☐

		Dollars	Cents
Income Attach Copy B of Form(s) W-2 here. Enclose but do not attach any payment with your return.	1 Total wages, salaries, and tips. This should be shown in box 1 of your W-2 form(s). Attach your W-2 form(s). 1		
	2 Taxable interest income. If the total is over $400, you cannot use Form 1040EZ. 2		
	3 Unemployment compensation (see page 9). 3		
	4 Add lines 1, 2, and 3. This is your **adjusted gross income.** If under $9,770, see page 9 to find out if you can claim the earned income credit on line 8a. 4		
Note: *You must check Yes or No.*	5 Can your parents (or someone else) claim you on their return? **Yes.** Enter amount from worksheet on back. ☐ **No.** If **single,** enter 6,800.00. If **married,** enter 12,200.00. See back for explanation. ☐ 5		
	6 Subtract line 5 from line 4. If line 5 is larger than line 4, enter 0. This is your **taxable income.** ▶ 6		
Payments and tax	7 Enter your Federal income tax withheld from box 2 of your W-2 form(s). 7		
	8a **Earned income credit** (see page 9). b Nontaxable earned income: enter type and amount below. Type ______ $ ______ 8a		
	9 Add lines 7 and 8a. These are your **total payments.** 9		
	10 **Tax.** Use the amount on **line 6** to find your tax in the tax table on pages 20–24 of the booklet. Then, enter the tax from the table on this line. 10		
Refund Have it directly deposited! See page 13 and fill in 11b, 11c, and 11d.	11a If line 9 is larger than line 10, subtract line 10 from line 9. This is your **refund.** 11a		
	▶ b Routing number		
	▶ c Type: Checking ☐ Savings ☐ d Account number		
Amount you owe	12 If line 10 is larger than line 9, subtract line 9 from line 10. This is the **amount you owe.** See page 13 for details on how to pay. 12		

I have read this return. Under penalties of perjury, I declare that to the best of my knowledge and belief, the return is true, correct, and accurately lists all amounts and sources of income I received during the tax year.

Sign here ▶ Keep copy for your records.
Your signature — Spouse's signature if joint return
Date — Your occupation — Date — Spouse's occupation

For Official Use Only

For Privacy Act and Paperwork Reduction Act Notice, see page 18. Cat. No. 14090A 1997 Form 1040EZ

Name ______________________________ Date ______________

38. State and Local Taxes

1. a. Buddy lives in a city that charges a 1% wage tax. His wages for the year were $19,876. How much wage tax does he owe?

   $______________

   b. The state in which Buddy lives charges an income tax of 3% of taxable income. His taxable income is $16,877. How much state income tax does he owe?

   $______________

   c. What is Buddy's combined city and state tax? $______________

2. a. Gert lives in a city that charges a 2% wage tax. Her wages for the year were $42,457. How much wage tax does she owe?

   $______________

   b. The state in which Gert lives charges an income tax of 4% of taxable income. Her taxable income is $31,578. How much state income tax does she owe?

   $______________

3. Sam lives in a state that charges 11% of all taxable income above $26,000. If Sam gets a raise that increases his taxable income from $26,000 to $28,000, how much of this increase will go toward state income tax?

   $______________

4. a. Jessica lives in a state that has a 15% inheritance tax. She has inherited $14,000 from her aunt. How much will the state take in inheritance taxes?

   $______________

   b. The lawyer who handled Jessica's inheritance charges 6% of the inheritance to do the legal work.

   How much will the lawyer charge? $______________

   c. How much will Jessica receive after state inheritance tax and lawyer's fees are deducted?

   $______________

5. a. Frank lives in a state that charges an inheritance tax of 15% on the first $10,000 of an inheritance and 7% on everything over $10,000. Frank inherited $14,000 from a grandparent. How much will the state take in inheritance taxes?

   $______________

   b. How much will Frank have left after the inheritance tax has been deducted? $______________

Name ______________________________ Date ______________

39. Use Tax

Your home state may have the right to collect use tax from you on any items you buy out of state that you will use in your home state. If you paid sales tax in another state on the merchandise you bought there, your home state (except Nevada) will allow you to deduct the sales tax you paid from the use tax that you owe. Ben lives in a state that charges a 6% use tax on all merchandise valued at more than $50.00. For some of the items Ben purchased, he paid sales tax to the state where his order was originated. Solve the following problems to see what use tax Ben paid during the past year.

1. Ben ordered a new computer hard drive, modem, and monitor from an out-of-state vendor for $724.00.

   What use tax does he owe his state? $____________

2. Last year, Ben also purchased some CD's by mail. He paid $46.85 for them. What use tax is owed?

   $____________

3. When Ben's refrigerator broke, he went to an appliance store in the next state over and bought a new one for $939.68. What use tax is due for this appliance? $____________

4. a. Ben paid a 4% sales tax on a $480.00 camera when he placed an order by phone. What amount of tax has he paid? $____________

   b. What use tax does he still owe his state for the camera? $____________

5. a. Ben joined a video club and ordered four videos for $19.95 each. At that time, he paid a 5% sales tax. What amount of sales tax did he pay to the state from which the videos were shipped?

   $____________

   b. How much use tax is owed to his state? $____________

6. a. Ben receives a 1½% discount on the amount of use tax due for outside repair and maintenance services. Ben was charged $565.00 to have his car repaired when it broke down on vacation—1,500 miles from home. What percent use tax is he required to pay?

   $____________

   b. What amount use tax is owed for the repair service? $____________

Name ______________________________ Date ______________

Section 7: Taxes Test

1. Caitlin bought two boxes of diskettes for $19.50 each. If the sales tax rate is 6%, how much did she spend on sales tax? $____________

2. How much is the sales tax on a bicycle that costs $239.00 in a state with a 5½% sales tax? $____________

3. How much more sales tax would you spend on a $46 item in a state whose sales tax is 7% compared to a state whose tax rate is 5%? $____________

4. On an income tax form, Tracy is asked to compute 28% of her earned income of $32,109.00. What is that amount? $____________

5. Brad has overpaid his federal income tax by $81.47 in taxes. If he only owed $1,006.33, how much was actually deducted from his wages? $____________

6. Samantha owes the federal government $515.79 in taxes. Her W-2 form shows that only $1,342.18 was actually deducted from her wages. What amount was listed on the tax table that should have been deducted from her wages? $____________

7. Sue inherited $9,000 from her aunt in a state that charges a 15% inheritance tax. How much will Sue receive after the inheritance tax is deducted? $____________

8. Mr. Winsett lives in a city that charges a 1½% wage tax. If his taxable income is $21,303, how much wage tax does he owe? $____________

9. Rolanda was charged a 5% use tax on a fax machine she purchased in a neighboring state. If the fax machine cost her $459.00, how much use tax did she pay her state? $____________

10. Diane ordered $117.95 worth of office supplies from a mail-order catalog. If her state is charging her a 6½% use tax, what amount does she owe? $____________

Section 8: Consumer Math Potpourri

TEACHER PAGES

40. Paying for Purchases

Computational Skills
Addition, Subtraction, Multiplication, Division

Mathematical Content
Percents/Proportions/Decimals, Currency

Procedure

1. Review with less advanced students questions 1, 4, and 8 before assigning this exercise.
2. For students who need additional practice in check writing, make additional copies of the exercise sheet and write these problems on the board:

   Paul bought a computer game for \$49.95 from the Ace Department Store. Make out the check for him, using today's date.

   Paul bought a CD for \$16.99 from Good's Record Shop. Make out the check for him using today's date.

1. \$19.72 (\$5.95 + \$3.99 + \$7.19 + \$2.59 = \$19.72)
2. \$3.11 (\$5.00 – \$1.89 = \$3.11)
3. \$70.05 (\$400.00 – \$329.95 = \$70.05)
4. \$14.05 (\$20 × 5 = \$100 –\$85.95 = \$14.05)
5. a. 4 bills (4 bills = 4 × \$50 = \$200)
   b. \$40.11 (\$200.00 – \$159.89 = \$40.11)
6. a. 9 bills (9 bills = 9 × \$20.00 = \$180.00)
   b. \$10.05 (\$180.00 – \$169.95 = \$10.05)
7. The check should be made out for \$36.68 (\$21.86 + \$15.00 = \$36.68). Be sure that the check is filled out completely.

41. Cost-Per-Unit: Liquid Measure

Computational Skills
Multiplication, Division

Mathematical Content
Currency, Weights/Measures

Procedure

1. You may want to bring in samples of liquid items and have students determine their unit cost.
2. Review with less advanced students questions 1 and 5 before assigning the exercise.

1. \$2.59 (\$3.89 ÷ 1.5 = \$2.59)
2. \$2.99 (\$2.99 ÷ 1 = \$2.99)
3. The 32-ounce size
4. \$.04 (\$.50 ÷ 12 = \$.04)
5. \$.02 (\$1.29 ÷ 67.6 = \$.02)

6. The 67.6 - ounce size
7. $.0009 ($.89 ÷ 946 = $.00094 = $.0009)
8. $.0007 (1.89 liters × 1,000 = 1,890 milliliters; $1.29 ÷ 1,890 = $.00068 = $.0007)
9. The 1.89 liter size
10. Answers will vary. Possible answers are:
    a. May be short of cash
    b. May only need small amount

42. Percent off

Computational Skills
Subtraction, Multiplication

Mathematical Content
Percents/Proportions/Decimals, Currency

Procedure

1. Point out that additional discounts are applied to the price after the first discount is taken. Do not add percent discounts first and then multiply to figure discount.
2. Review with less advanced students questions 1, 4, and 7 before assigning this exercise.
    1. $359.20 ($449.00 × .20 = $89.80; $449.00 – 89.80 = $359.20)
    2. $207.29 ($287.90 × .20 = $57.58; $287.90 – 57.58 = $230.32; $230.32 × .10 = $23.03; $230.32 – $23.03 = $207.29)
    3. $341.40 ($455.20 × .25 = $113.80; $455.20 – 113.80 = $341.40)
    4. $482.00 ($357.04 × 2 = $714.08; $714.08 × .25 = $178.52; $714.08 – $178.52 = $535.56 × .10 = $53.56; $535.56 – $53.56 = $482.00)
    5. $4.43 ($88.69 × .05 = $4.43)
    6. $9.86 ($197.25 × .05 = $9.86)

43. Magazine Subscriptions

Computational Skills
Subtraction, Multiplication, Division

Mathematical Content
Currency

Procedure

1. Review with less advanced students questions 1 and 2 before assigning Part I.
2. Review with less advanced students questions 7 and 8 before assigning Part II.

Part I

1. \$50.70 (\$1.95 × 26 = \$50.70)
2. \$21.58 (\$1.12 × 26 = \$29.12; \$50.70 – 29.12 = \$21.58)
3. \$9.71 (\$29.12 ÷ 3 = \$9.706)
4. \$101.40 (\$1.95 × 52 = \$101.40)
5. \$43.16 (\$1.12 × 52 = \$58.24; \$101.40 – 58.24 = \$4;.16)
6. \$19.41 (\$58.24 ÷ 3 = \$19.41)

Part II

7. \$27.00 (\$2.25 × 12 = \$27.00)
8. \$1.25 (\$15 ÷ 12 = \$1.25)
9. \$12.00 (\$27.00 – 15 = \$12.00)
10. \$1.13 (\$27.00 ÷ 24 = \$1.13)
11. \$1.06 (\$38.00 ÷ 36 = \$1.06)
12. \$7.00 (\$15 × 3 = \$45; \$45 – 38 = 7)

44. Mail-Order Form

Computational Skills

Addition, Subtraction, Multiplication

Mathematical Content

Percents/Proportions/Decimals, Currency

Procedure

1. You may want to bring in catalogs and have students look through and place mock orders.
2. Work through the entire exercise with students of varying abilities.
3. To provide additional practice for less advanced students, make additional copies of the exercise and write these instructions on the board:

   Sally Reiner wants to order for herself one North Star cherry tree, a packet of bush bean seeds, and a packet of Shasta Daisy seeds.

   She wants to send a fern to Mae Sloan at 1333 River Drive, Crystal, CA 90131

   Fill out the order form for her.

   Note: For the practice exercises, the total of front side equals \$42.53; the total of back side equals \$9.99; the subtotal equals \$52.52; CA sales tax equals \$2.47 (.06 × \$41.24); total amount equals \$59.99.

FRONT SIDE:

Special instructions for collections and seeds: Enter in "Letter Code" column the boxed letter (for example, C) following the catalog number. Enter in "How Many" column the number of units you want.

SHIP THESE ITEMS TO ME AT THE ABOVE ADDRESS (See other side for gifts shipped to my friends and relatives)

Catalog No.	Letter Code	How Many	Name of Item	Page No.	Price	
1807—7		6	Fern (2 x $25.00)	21	50	00
2904—4	7	1	Blue Lake Bush Beans	54	4	00
3066—5	A	2	Shasta Daisy	32	1	99
—						

Total above	55	99
Total from other side	39	95
Sub-total	95	94
Handling charge	5	00
CA sales tax: 6% except vegetable seeds: (.06) (50 + 1.99 + 39.95)= (.06) (91.94)= 5.52	5	52
Total amount – check, credit card, or money order enclosed (Please do not send cash or stamps) (95.94+5.00+5.52=106.46)	106	46

BACK SIDE:

SHIP THESE ITEMS AS GIFTS TO MY FRIENDS
(If no address is given, ship items to my address)

	Catalog No.	Letter	How Many	Name of Item	Page No.	Price	
Ship to Marty Driezinger	1048—8		1	North Star	16	39	95
Address 112 Lane St.	—						
City/State Spring City, CA Zip 90028	—						
Sign card from Donald White	—				TOTAL	39	95

Note: Subtotal minus cost of vegetable seeds = $95.94 – $4.00 = $91.94 = taxable amount

45. Compact Disc and Book Clubs

Computational Skills

Addition, Multiplication, Division

Mathematical Content

Currency

Procedure

Review with less advanced students questions 1a, 1b, 4a, 4b, and 4c before assigning the exercise.

Part I

1. \$144.91 (\$1.00 + 9 × \$15.99 = \$1.00 + \$143.91 = \$144.91)
2. a. \$350.75 (\$1.00 + 25 × 13.99 = \$1.00 + 349.75 = \$350.75)
   b. \$11.69 (\$350.75 ÷ 30 = \$11.69)
3. Answers will vary

Part II

4. a. \$56.24 (\$14.00 + 19.89 + 22.35 = \$56.24)
   b. \$11.39 (\$3.00 + \$15.90 + 18.75 + 22.50 + 19.55 = \$79.70; \$79.70 ÷ 7 = \$11.39)
5. \$13.15 (\$21.50 + 17.95 + 16.95 + 22.50 + 15.97 + 20.50 = \$115.37; \$115.37 + 3.00 = \$118.37; \$118.37 ÷ 9 = \$13.15)

46. Advertising Costs

Computational Skills

Addition, Subtraction, Multiplication, Division

Mathematical Content

Tables/Graphs, Whole Numbers, Currency

Procedure

1. Discuss with students the importance of advertising. List different methods of advertising that exist and which methods are feasible to use when trying to promote a small business. Which of the methods discussed are the most effective?
2. Go over problem 2 as a review with the class. Point out that it is not realistic to have a fraction of an advertisement. Therefore, the answer is rounded down from 6.667 to 6. If you were to round up, this would exceed the allowable budget.
3. Discuss the importance of budgeting costs. After students complete the exercise, point out that in problem 5, the situation presented is not feasible for Alex to use since the amount goes over her allowable budget.

1. 2 (\$500 ÷ \$250 = 2)
2. 6 (\$500 ÷ \$75 = \$6.67 = 6)
3. 5 (\$500 ÷ \$100 = 5)
4. 20 weeks (\$500 ÷ \$25 = 20)
5. \$625 (\$25 × 4 = \$100; \$75 × 7 = \$525 + \$100 = \$625)
6. \$500 (\$75 + \$25 × 7 + \$250 = \$500)
7. \$130 (\$250 + \$120 = \$370; \$500 – \$370 = \$130)
8. a. yes (5 × 75 = 375; 25 × 5 = 125; 500 – 375 – 125 = 0)
   b. \$500—her entire budget
9. 1 space ad (\$250), 1 radio spot (\$75), 1 Internet ad without photo (\$100), 1 week classified ad (\$25)

Section 8 Test

1. $1.01 (5 × $20 = $100 – $98.00 = $1.01)
2. 1 liter (1.04 ÷ 4 = .26)
3. $27.00 ($179.99 × .15 = $27.00)
4. $43.18 ($29.99 × 2 = $59.98 × .2 = $12.00; $59.98 – $12.00 = $47.98 × .1 = $4.80; $47.98 – $4.80 = $43.18)
5. $0.42 ($19 ÷ 12 $1.58; $2.00 – $1.58 = $0.42)
6. $100 ($25 × 4 = $100)
7. $8.28 (6 × $1.29 = $7.74 × .07 = $0.54 + $7.74 = $8.28)
8. $11.88 (6 × $16.99 = $101.94 + $4.99 = $106.93 ÷ 9 = $11.88)
9. $43.50 ($14.50 × 3 = $43.50)
10. $2.29 (80 ÷ 35 = $2.29)

Name ______________________ Date ______________

40. Paying for Purchases

Answer the following questions.

1. Mr. Lang wanted to buy a steak for $5.95, a box of laundry detergent for $3.99, a jar of coffee for $7.19, and a jar of mayonnaise for $2.59. What was the total cost of these items. $__________
2. Jesse bought a can of oil for his motorcycle. It cost $1.89. He gave the cashier $5.00. How much change should he have received? $__________
3. Jane bought a fax machine for her home office. It cost $329.95 and she paid by check. To get cash back, she made the check for $400.00. How much money should she receive as change? $__________
4. Ruth bought a tennis racket. It cost $85.95. She gave the cashier five 20-dollar bills. How much change should she have received? $__________
5. a. Sam bought a portable compact disc player. It cost $159.89. He had seven 50-dollar bills in his wallet. How many bills should he have given to the cashier? __________

   b. How much change should Sam have received? $__________
6. a. Julien bought an AM-FM cassette radio. It cost $169.95. He had ten 20-dollar bills in his wallet. How many bills should he have given to the cashier? __________

   b. How much change should Julien have received? $__________
7. Paul Raid went to ABC Supermarket. His total bill was $21.68. He wanted to pay with a check and get $15 in change. Fill out Paul's check for him using today's date. (*Hint:* Paul's check should be made out for $15 more than the amount of the bill.)

CALIFORNIA BANK

West Office
373 Thousand Oaks Boulevard
Thousand Oaks, California 91360

90-301
122

442

______________________ 19 ____

Pay to the order of ______________________ **$** ________

______________________ **dollars**

Paul Raid
229 North Street
Thousand Oaks, CA 91360

____

⑆1222⑈301 26520416 044

Name ______________________ Date ______________

41. Cost-Per-Unit: Liquid Measure

Read the information in the boxes and answer the questions.

> The following information is for two sizes of Brand D cooking oil:
>
> A 48-ounce (1.5 quart) bottle costs $3.89.
> A 32-ounce (1 quart) bottle costs $2.99.

1. What is the cost per quart if you buy the 48-ounce size? $____________
2. What is the cost per quart if you buy the 32-ounce size? $____________
3. Which size costs more per ounce? ____________

> The following information is for two sizes of Brand C soda:
>
> A 355-milliliter (12-ounce) can costs $.50.
> A 2,000-milliliter (67.6-ounce) bottle costs $1.29.

4. How much does the 12-ounce can cost per ounce? (Compute your answer to three decimal places and round to two places.) $____________
5. How much does the 67.6-ounce bottle cost per ounce? (Compute your answer to five decimal places and round to four places.) $____________
6. Which size is the better buy? ____________

> The following information is for two sizes of Brand E milk:
>
> A 946-milliliter (1 quart) bottle costs $.89.
> A 1.89-liter (half gallon) bottle costs $1.29.

Hint: To convert liters to milliliters, multiply the number of liters by 1,000.

7. How much does the 946-milliliter bottle cost per milliliter? (Compute your answer to five decimal places and round to four places.) $____________
8. How much does the 1.89-liter bottle cost per milliliter? (Compute your answer to five decimal places and round to four places.) $____________
9. Which size bottle of milk is the better buy? ____________
10. In each case, why might you buy the size that is *not* the better buy? ____________

____

Name ______________________________ Date ______________

42. Percent Off

In the boxes are the percents you can save by buying items on sale. Answer the questions under each box.

Hint: To find 20% of an amount, multiply by .20.

Leather Coats and Jackets

Price CUT 20%

1. The leather coat you want normally costs $449.00. How much will it cost while it is on sale? $__________
2. This store is holding a special holiday sale. If you purchase any jacket during the first two hours the store is open, an additional 10% discount is taken off the already discounted price. The leather jacket you want normally costs $287.90. How much will it cost on sale during the first two hours the store is open?

   $__________

25% OFF

Custom Blinds

3. Custom blinds for your living room normally cost $455.20. How much will they cost while they are on

   sale? $__________
4. Custom blinds for your dining room normally cost $357.04. If you purchase two sets, you get an additional 10% off the sale price of each set. How much will two sets cost while they are on sale?

   $__________

Save 5% Order Early!

Send us your order by September 15 and deduct 5% from the total value of your order. Get all your holiday shopping done early (and easily).

5. The holiday presents you want to order cost $88.69. How much will you save if you order before

   September 15? $__________
6. The holiday presents Ms. Perez wants to order cost $197.25. How much will she save if she orders

   before September 15? $__________

Name ________________________________ Date ________________

43. Magazine Subscriptions

Part I

In the box is an ad offering a subscription to a weekly magazine. The "cover price" is the price you would pay at a newsstand. Read the ad and answer the questions.

Over 40% off the $1.95 cover price. That's just $1.12 an issue–delivery to your home is FREE.

Send me: ❑ 18 months (78 issues) ❑ 9 months (39 issues)
❑ 1 year (52 issues) ❑ 6 months (26 issues)

OVER 40% OFF the cover price!

NAME ________________________________

ADDRESS ________________________________

CITY ________________ STATE ________________ ZIP ________________

❑ Payment enclosed. ❑ Bill me later. ❑ Bill me in three monthly installments.

1. What is the total cost of 26 issues at the cover price? $____________
2. How much would you save if you subscribed for 26 issues instead of buying 26 issues at the cover price? $____________
3. If you chose to be billed in installments for a six-month subscription, how much would each installment be? $____________
4. What is the total cost of 52 issues at the cover price? $____________
5. How much would you save if you subscribed for 52 issues instead of buying 52 issues at the cover price? $____________
6. If you chose to be billed in installments for a one-year subscription, how much would each installment be? $____________

(continued)

Name ______________________ Date ______________

43. Magazine Subscriptions *(continued)*

Part II

In the box are the subscription rates for a monthly magazine. Each issue costs $2.25 on the newsstand. Read the ad and answer the questions.

One Year	**$15**
Two Years	**$27**
Three Years	**$38**

7. What is the total cost of 12 issues at the newsstand price? $____________
8. How much does each issue cost if you subscribe for one year? $____________
9. How much would you save if you subscribed for one year instead of buying 12 issues at the newsstand?

   $____________
10. How much does each issue cost if you subscribe for two years? $____________
11. How much does each issue cost if you subscribe for three years? $____________
12. How much would you save if you subscribed for three years instead of buying three separate one-year subscriptions? $____________

Name ______________________________ Date ______________

44. Mail-Order Form

Described in the boxes below are plants, trees, and seeds Donald White wants to order through the mail. The front and back sides of the order form are in the boxes on the next page. Fill out the order form.

Mr. White wants to order six of these:

> 1807-7 FERN. Popular plant for hanging baskets. Each $9.99, 3 for $25.00.
> **Page 21**

Mr. White wants to order one-half pound of these:

> 2904-4 BLUE LAKE BUSH BEANS. Delicious, plump, round. (Seeds) [A] Pkt $1.29 [T] ½ lb. $4.00
> **Page 54**

Mr. White wants to have one of these sent to his friend Marty Driezinger who lives at 112 Lane Street, Spring City, CA 90028:

> 1048-8 NORTH STAR. Dwarf 8 ft. trees bear large cherries. Each $39.95; 2 or more, each $34.95.
> **Page 16**

Mr. White wants to order two packets of these:

> 3066-5 SHASTA DAISY. Single flowers 4 to 5 inches across. Yellow centers. (Seeds) [A] Pkt. $1.29; 2 – $1.99 **Page 54**

Hints: Letter codes are letters in small boxes. Not all items have letter codes. To compute the sales tax, subtract the cost of vegetable seeds from the subtotal and multiply the difference by .06.

(continued)

Name ______________________________ Date ______________

44. Mail-Order Form *(continued)*

FRONT SIDE:

Special instructions for collections and seeds: Enter in "Letter Code" column the boxed letter (for example, [C]) following the catalog number. Enter in "How Many" column the number of units you want.

SHIP THESE ITEMS TO ME AT THE ABOVE ADDRESS (See other side for gifts shipped to my friends and relatives)

Catalog No.	Letter Code	How Many	Name of Item	Page No.	Price	
—						
—						
—						
—						

Total above		
Total from other side		
Subtotal		
Handling charge	*5*	*00*
CA sales tax: 6% except vegetable seeds:		
Total amount – check, credit card, or money order enclosed (Please do not send cash or stamps)		

BACK SIDE:

SHIP THESE ITEMS AS GIFTS TO MY FRIENDS
(If no address is given, ship items to my address)

	Catalog No.	Letter	How Many	Name of Item	Page No.	Price
Ship to	—					
Address	—					
City/State Zip	—					
Sign card from	—				TOTAL	

Name ______________________________ Date ____________________

45. Compact Disc and Book Clubs

Part I

Answer these questions about a compact disc (CD) club. When you divide, compute your answers to three decimal places and round off to two.

1. Jan joined a compact disc club. When she joined, she sent in $1.00 and received five CD's. She bought nine more CD's during the next three years for $15.99 each, which she agreed to do when she joined. How much did the 14 CD's she received cost her?

   $______________

2. a. Sam joined the same club that Jan did. He received five CD's for $1.00 when he joined. During the next three years, he bought 25 more CD's at $13.99 each. How much did the 30 CD's he received cost him?

   $______________

   b. How much did Sam pay per CD for the 30 total he received? $______________

3. In the box is part of the description of the compact disc club. Read it and name an advantage or disadvantage of belonging to the club. ______________________________

> Every four weeks (13 times a year) you'll receive the Club's music magazine, which describes the Selection of the Month for each musical interest . . . plus hundreds of alternates from every field of music. In addition, up to six times a year you may receive offers of Special Selections, usually at a discount off regular prices.
>
> If you wish to receive the Selection of the Month or the Special Selection, you need do nothing—it will be shipped automatically. If you prefer an alternate selection, or none at all, simply fill in the response card always provided and mail it by the date specified.

Part II

Answer these questions about a book club.

4. a. Tori joined a book club. She sent in $3.00 and received three books valued by the club at $14.00, $19.89, and $22.35. What is the total value of the books Tori received for $3.00?

   $______________

   b. When she joined, Tori agreed to buy four more books at regular prices. The books she bought cost $15.90, $18.75, $22.50, and $19.55. How much did Tori pay per book for the seven books she received?

   $______________

5. Jacob joined the same book club. He sent in $3.00 and received books valued by the club at $16.50, $34.75, and $22.89. Jacob bought six more books at these prices: $21.50, $17.95, $16.95, $22.50, $15.97, $20.50. How much did he pay per book for the total of nine books he received?

   $______________

Name ______________________________ Date ______________

46. Advertising Costs

Alex is starting a computer consulting business. She has put aside $500 for advertising costs associated with promoting her services. She has researched some of the most effective advertising methods and their costs. A summary of her findings is outlined in the table below. Use the information to answer the following questions.

Method	*Cost*	*Terms*
Radio spot	$75	30-second announcement
Internet server	$100 $120	up to 200 words (no photo) for 90 days up to 200 words (with photo) for 90 days
Classified ad	$25	20 words per week run of ad
Space ad	$250	1/8 page in Sunday edition of local paper

1. If Alex wants to use all her advertising funds to run a ⅛-page ad, how many ads could she take out?

   ____

2. If she uses her funds to buy radio airtime only, how many commercials could she afford?

   ____

3. If she advertised solely on the Internet, how many days could she run an ad with no photo?

   ____

4. If Alex decides to use $500 to place one classified ad, what is the maximum number of weeks her ad would run? ______________

5. If Alex chooses to run a 20-word classified ad for four weeks and play a daily radio commercial for an entire week, what would this cost her? $______________

6. What are Alex's costs to place one 30-second radio ad, a one-week classified ad, and one space ad in the Sunday paper? $______________

7. If a space ad is placed in the Sunday paper and an ad containing a photo is placed on the Internet for 90 days, what amount of Alex's budget would be left over? $______________

8. a. Does Alex have enough money in her budget to run five radio spots and put a classified ad in for a week? ______________

   b. What would the total cost be? $______________

9. If Alex advertises her business using all four methods described while still staying within her budget, how many of each type of ad can she take out and for what length of time? ______________________________

   ____

Name ______________________________ Date ______________

Section 8: Consumer Math Potpourri Test

1. Le ys bought a portable CD player for $98.99. He gave the cashier five 20-dollar bills. How much change should he have received? $__________

2. A four-liter bottle of spring water costs $1.04. A one-liter bottle costs $0.37. What size costs the most per liter? $__________

3. A store is advertising that all winter coats will be sold for 15% off the regular price during March. The coat you want regularly costs $179.99. How much will you save if you purchase it during the month of March? $__________

4. Jeans are currently on sale for 20% off the regular price of $29.99. If you purchase two pairs, you can get an additional 10% off the sale price. What would the cost of two pairs of jeans be? $__________

5. You can subscribe to a monthly magazine for one year for $19.00. Each issue costs $2.00 on the newsstand. How much would you save *per issue* if you subscribed for a year instead of buying the magazine at the newsstand? $__________

6. A nursery is advertising rosebushes at $14.95 each or two for $25.00. How much would eight rosebushes cost you? $__________

7. Ed bought some flower seeds from a catalog. The price per package of seeds was $1.29. What is the total amount Ed paid for six packages of seeds including a 7% sales tax? $__________

8. Carrie joined a compact disc club. She sent in $4.99 and received three CD's and agreed to purchase six more CD's during the year at the regular price of $16.99 each. How much will Carrie spend per CD for the nine CD's she will receive this year? $__________

9. Maxine placed an ad in the classified section to try and sell her old baseball card collection. The ad costs $14.50 per week for a four-line ad. How much will it cost Maxine to run a four-line ad for three weeks? $__________

10. Steven wants to put an ad over the Internet to sell his car. One Internet advertising service charges $80 to place an ad for five weeks. If Steven advertises though this service, what will the ad cost him per day? $__________